The Greening of Socialism

The Greening of Socialism

Climate Change and Marx in the 21st Century

Sanjeev Ghotge

LEXINGTON BOOKS
Lanham • Boulder • New York • London

Published by Lexington Books
An imprint of The Rowman & Littlefield Publishing Group, Inc.
4501 Forbes Boulevard, Suite 200, Lanham, Maryland 20706
www.rowman.com

6 Tinworth Street, London SE11 5AL, United Kingdom

British Library Cataloguing in Publication Information Available

Library of Congress Cataloging-in-Publication Data

Names: Ghotge, Sanjeev, author.
Title: The greening of socialism : climate change and Marx in the 21st century / Sanjeev Ghotge.
Description: Lanham : Lexington Books, [2020] | Includes bibliographical references and index. | Summary: "Three existential crises confront global humanity: climate change, the environmental crisis and the arms race. This book argues their resolution will be through the development of a green socialism"-- Provided by publisher.
Identifiers: LCCN 2020002145 (print) | LCCN 2020002146 (ebook) | ISBN 9781498595735 (cloth) | ISBN 9781498595742 (epub)
Subjects: LCSH: Communism and ecology. | Climatic changes--Economic aspects. | Climatic changes--Social aspects. | Sustainable development.
Classification: LCC HX550.E25 G75 2020 (print) | LCC HX550.E25 (ebook) | DDC 335.4--dc23
LC record available at https://lccn.loc.gov/2020002145
LC ebook record available at https://lccn.loc.gov/2020002146

Dedicated to the next generation . . . and beyond?

I could a tale unfold whose lightest word

Would harrow up thy soul, freeze thy young blood,

Make thy two eyes, like stars, start from their spheres,

Thy knotted and combined locks to part,

And each particular hair to stand on end,

Like quills upon the fretful porpentine.

—Shakespeare, writing about climate change in *Hamlet*

Contents

Preface

Back and Forth

This monograph has emerged out of a series of coincidences. During 2017, in the course of correspondence with Salvatore Engel di Mauro (Saed), the editor of the journal *Capitalism Nature Socialism*, I had been urging him to devote a special issue to Marx and his continuing relevance in the twenty-first century, especially in the light of climate change for which capitalist society has no relevant answers—not only that, capitalist society itself was the greatest single culprit in both climate change and the global environmental crisis. Not only did capitalist society not possess the answers to these twin crises threatening humanity, it did not even possess the capacity to recognize its own culpability, pass judgment on itself, and wish itself out of existence if humanity was to be saved. Saed responded by throwing the ball back in my court, inviting me to write a piece with a promise of publishing it in 2018, the 200th anniversary of Marx's birth. That invitation resulted in a two-part essay, published in Volume 29 Numbers 2 and 3 of *CNS* in 2018; that essay, in turn, resulted in an invitation from the present publishers to expand the original essay into a monograph.

Capitalist society would have wished that Marx had never been born; the next best tactic has been to very largely ignore his contributions to a political understanding of the human condition on earth even two hundred years after his birth. Marx's *Communist Manifesto* apparently has sold more copies around the world than any other publication, second only to the Bible. By the yardstick of capitalism, that would make Marx a huge best-selling author which capitalist society would like not merely to deny but to suppress as well. Yet, when an author is banned, even more people sit up and take note of him or her and decide to read him or her in order to understand what it is that is considered objectionable, so that they can make up their own minds about the matter and its author. This has kept Marx's writing alive in generation after generation over the last 150 years; that itself should be taken as a testament to the continuing vitality of his thought and the vigor of his writings. The younger generation would be well advised to pay heed to his writings in the original.

There is, however, a deeper ground for writing, one that is owed to the next generation and future generations. Climate change will ravage the earth for generations and centuries to come and the younger generation must be made aware of what is to follow, so that they have a chance to make a choice of how the future is to be lived. There is a moment of existential freedom available to a generation as they near the end of their educational life and step into the stream. For that brief moment, the chance to choose their life remains open: will foreknowledge of the impending future help them choose a wiser course for their lives and to create a better collective future for themselves and others?

One road leads to a barren and devastated future, the other to a greener and more humanly hopeful and fulfilling alternative. The choice will be fateful.

After completing writing the monograph, like all bad authors I have revisited the beginning and wanted to rewrite it.

I have resisted the temptation, limiting myself to renaming it "Back and Forth."

Chapter One

Introduction and Overview

This monograph addresses several interlinked themes: the multiplicity of existential crises confronting global humanity in the twenty-first century and beyond; the rootedness of these crises as inherently lying in the very fabric and operations of the globalized capitalist system and its structure of production and the class interests bearing on its continuance; and lastly, the possible solutions to these crises if we remove the blinkers of capitalist class interests that are currently preventing the search for solutions transcending the interests of the capitalist system.

Since the above analyses require a broad framework of political economy if the main outlines of the problems and their eventual resolution or transcendence has to be rendered visible, given that global capitalism resembles an octopus, we have chosen to use the critique of the capitalist system by Marx as an underlying theoretical basis. This enables us to hack through the thicket of deception and self-deception that the capitalist system has thrown up in self-defense and self-justification over the last couple of centuries, an expert play of smoke and mirrors. However, such an approach inevitably assumes some basic understanding of Marx as a philosopher and a social scientist who combined theory with practice in developing his profound and critical understanding of capitalism; therefore, in using and applying concepts from his theoretical framework we choose to accord him the respect that he richly deserves for his lifelong labors. Marx belongs to humanity at large and his writings are relevant to the immediate past as also the present of Europe, Asia, Africa, and North and South America as well as the countries of the socialist and communist bloc. More to the point, his writings are relevant to the future as well. Therefore, any discussion on Marx in the following pages is neither accidental nor hagiographic, it is central to throwing light upon the present and future condition of the world and a serious consideration of saving it from self-destruction. Of necessity, such an analysis will neither be obsequious to capitalist interests nor subservient to them. The truth about the present condition of the world as we find it, and the likely future toward which it is now headed, is not well served by such dubious values. Whatever needs to be truthfully stated will be so stated, bluntly, subject to the constraints imposed by good science, enlightened theory, and "common sense" sensus communis (Arendt 1971, pp. 51–54).

The three major crises confronting global humanity, for the purposes of this monograph, are the climate crisis, the global environmental crisis, and the nuclear arms crisis, each of which is independently capable of destroying life on earth or life as we know it. It is not the intention of this monograph to propose that these are the only three life-threatening "existential crises" wherein the term "existential" is used to denote the very existence of life on earth, both human and all other species. Beyond these three, there seems to be at least one other crisis brewing at the microbiological level, which could potentially emerge as an "existential" crisis. This is a crisis caused by the emergence of antibiotic-resistant microbes and viruses emerging from either collapsing ecological habitats or human economic practices such as industrialized animal husbandry. Although one can see the links of this crisis with the massive globalization of economic activities under a generalized capitalist model as well as possibly with the beginnings of climate change effects, it is beyond the capability of this author to penetrate the thickets of the discourse on biology, ecology, human and animal medicine, and medical practices and pharmacology, these being the disciplines that will have to shed light on how to deal with this particular crisis. For the purposes of this monograph, we have enough on our plate in outlining the three major crises indicated above, that is, climate, environment, and the nuclear armaments race. In terms of urgency, the order could possibly be reversed, that is, the nuclear arms race could destroy human civilization and human life on earth in a matter of days or years ("the nuclear winter"); the environmental crisis which is a single phrase covering the impact of human activity on oceans, forests, life forms, water, land, and so on which could slowly choke human existence over centuries of drowning in our own filth; and the climate crisis which threatens to wipe out most life on the earth in a matter of a few centuries ("the sixth mass extinction," which could include human extinction). However, if time is temporarily set aside as the variable for judging urgency, then the climate crisis is the one that is on track to ensure human extinction or near extinction. Since we at the start of the twenty-first century are beginning to see the tip of the iceberg, a significant part of this monograph will be devoted to this crisis. Moreover, by now there is adequate material on record concerning the other two crises; the same cannot yet be said about the climate crisis, wherein we are refusing to connect the dots. For the purposes of this monograph, may we suggest that these three crises be respectively referred to as "Fast Death," "Slow Death," and "Certain Death" of humanity?

The second part of the monograph concerns the linkages of these crises with the global structure of the political economy under capitalism: for this part of the analysis, the understanding provided by Marx's analysis of the political economy of capitalism provides a suitable foundation both for critiquing the current global paradigms and for pushing the analytical envelope toward a greener, more socialistic global pattern of society which may be able to overcome the three major crises now confronting humanity. For the purpose of condensing the argument, this part of the examination will critique four broad epistemological models of political economy, namely the Hayekian model, the Keynesian model, the State Socialist model, and the Green Keynesian model. This, the critical part of this monograph, will clear the way for proposing a Green Socialist model as an alternative global future as distinct from the other four models. To an extent, this will also indicate the relevance of Marx's thought on political economy as providing extremely valuable insights into the genesis of the crises and therefore the need to study that important stream of thought in the entirety of its implications without getting trapped into marxist exegesis (that is, the bookish quotation of Marx in the manner of the devil quoting scripture).

The third and final part of this monograph outlines and argues the case for a fifth model of political economy—the Green Socialist model. Like the other models of political economy, this is a condensed version for the purpose of preserving the clarity of argument and for indicating that a paradigm shift in societal thinking is urgently required in the face of the dire existential crises confronting humanity. The model will try to indicate that the financial, technological, and human "resources" are already available within the global system but that they cannot be energized under the regime of global capital that is now the dominant prevailing mode. While the political "ends" are increasingly becoming clearer as we rush headlong toward the cliff, the political "means" for achieving those ends remain riven with a lack of clarity as regards the mode and agency for change. The model therefore limits the analysis to delineating the necessary building blocks for the construction of a Green Socialist economy but prefers to leave the choice of means of transformation to agencies at the level of every nation state. This is not an attempt to sidestep the questions of "revolution," rather it pays due heed to Engels' insight that "revolutions are born out of helplessness." Throughout the analysis, we also try to maintain the spirit of Marx's assertion that "men make their own history but not necessarily under conditions of their own choosing."

In effect, the crisis in human affairs may be forcing such a choice on humanity in the near future.

Chapter Two

200 Years after His Birth

Appreciating the Relevance of Marx in the Twenty-First Century

Marx was born in 1818 at Trier, now in Germany. His writings span the period from 1844 to his death in 1883. At the time of his death, he was still at work on his magnum opus, *Capital* (*Das Kapital*) of which the first volume had been published in German and the subsequent volumes were compiled and produced by his lifelong friend and collaborator, Friedrich Engels. Marx was a highly learned man; by today's fashion he should be called Dr. Marx since he had what would be considered a double PhD—in philosophy and jurisprudence. This certification enabled him to fight his own cases in courts across several continental European countries and in England where he finally took up residence after being expelled from three European countries, Germany, France, and Belgium, on account of his "inflammatory" writings. His political enemies never stopped persecuting him and attempting to block his writings, which would broadly fall into two categories or periods (1) 1844 to 1871 and (2) 1875 to 1883, the latter period devoted to *Das Kapital*. However, it needs to be noted that *Das Kapital* picked up the threads of his own arguments, first outlined in "The Economic and Philosophical Manuscripts" (1844) and later broadened in "Grundrisse—The Foundation and Critique of Political Economy" (1867). There is an amazing continuity in *Das Kapital* across nearly forty years of observation and participation in the political life of kingdoms in Western Europe, given the dislocations and persecutions to which he was subjected throughout his life. Both political activists and academics across the world would perhaps appreciate the lifelong pressures under which he worked and lived to produce and leave behind a most remarkable corpus of work, viewed in its entirety.

The periodization of his writings from 1844 to 1871 and 1875 to 1883 are not arbitrary; they reflect major event structures in European history as well as Marx's involvement in those events. In a nutshell, the period from 1844 to 1871 corresponded to a sweep of "national revolutions" across Western, Southern, and Central Europe. Germany was forged in 1871 out of about 320 feudatory states, kingdoms, and duchies; Italy also was formed in 1871 out of a similar messy agglomeration of papal states, kingdoms, and independent cities and parts of the Austro-Hungarian Empire. The emergence of both of these nation-states marked the time for the eventual demise of the Austro-Hungarian empire straddling Western and Central Europe. The year 1871 also saw a Europe dominated by seven empires: the British, the French, the Austro-Hungarian (Habsburg), the German (Second Reich), the Italian, the Russian Czarist, and the Turkish Ottoman. While the political aspirations and rebellions of the underclasses were successfully crushed across all these lands nevertheless a substrate of socialist consciousness emerged and remained across all these territories, the first time in world history. This proletarian consciousness marked itself as distinct from the worldview of both the dominant classes: the conservative landed classes of the earlier feudal regimes and the liberal capitalist classes that created the basis of their wealth and power through the ownership of financial capital, or industrial capital invested in factory-based production or trade capital making profits through processes of mercantile exchange. It was equally in the class interest of the capitalist classes as a whole to acquire foreign territories as colonial possessions of their respective empires, to feed resources and labor into their factories, to provide captive markets for their surplus goods, and to absorb their financial capital as investment, all under the aegis of their respective empires. It was the genius of Marx that was able to cut through the hypocrisies of his times to reveal the underlying combination of class interests that propelled events globally. This global vision of Marx, rooted in the universal vision of the French Revolution, is important to deal with the three global problems initially outlined: nuclear armaments and war, the environmental crisis, and the climate crisis.

The Europe that Marx saw as a young man cannot be understood without understanding the enormous political impact of the French Revolution on the rest of continental Europe. The French Revolution toppled the Ancien Régime of France in its entirety—royalty, aristocracy, and ecclesiastical power—to be replaced by a Republic in which the people governed themselves according to their own constitution. The old feudal powers along with all their decrepit institutions—legal, political, military, and economic—were swept aside in one historic upsurge of popular anger which sent shock waves across continental Europe, still under feudal dynastic rule. The confrontation here is best expressed in the words of a modern British historian:

> It was, in modern terms, the issue of whether two forms of society based on totally different principles, could peacefully co-exist. France within her own territories had ended feudalism, destroyed the pretensions of royal absolutism, and founded new institutions on the principles of sovereignty of the people and personal liberty and equality. The old institutions, which had been overthrown in France, remained established in her continental neighbors. The influence of the Revolution was spreading, undermining the position of other rulers and implicitly challenging the survival of serfdom, feudalism and absolutism everywhere. The revolutionary ideals were too dynamic to be ignored by the established order. (Thomson 1974, p. 15)

These two "forms of society"—republican versus feudal dynastic—could not co-exist in Europe, resulting in an alliance of the feudal—dynastic powers of old Europe (Austrian–Habsburg, Prussian, Dutch, Spanish, and an émigré army raised by the deposed aristocracy of France) and Britain which attacked the nascent republic of France. The republican armies, under the leadership of Napoleon, then a general, were largely victorious and Napoleon used his personal popularity in France to usurp power and declare himself emperor. From 1803 to 1811, Napoleon won a series of victories across the European continent all the way up to successfully invading Moscow before the combined strength of his opponents, now joined by Russia, overwhelmed and crushed him in 2013. Yet, even during the relatively brief period of French ascendancy across Europe, republican ideas were forcefully implemented across occupied and dependent territories in Europe, in accordance with a revolutionary manifesto of November 1792:

> In December, the [French] Assembly served notice on Europe that France would enforce revolutionary social principles everywhere. Occupation of any territory by French armies, they declared, would be accompanied by the ending of feudal obligations and the confiscation of clerical and aristocratic property. (Thomson 1974, p. 15)
>
> Napoleon extended and perpetuated the effects of the French Revolution in Europe by destroying feudalism in the Low Countries, in much of Germany, and in Italy. Feudalism as a legal system, involving noble jurisdiction over peasants, was ended; feudalism as an economic system, involving payment of feudal dues by peasants to nobles, was ended. . . . The claims of the church were never allowed to stand in the way of this reorganization. Middle classes and peasants became, like nobles, subjects of the State, all equally liable to pay taxes. (Thomson 1974, p. 48)

Thus did the fates of revolution and war get intertwined across Europe.

> The sweeping successes of the first revolutionary armies become understandable only if the greedy and separatist characteristic of their opponents is realized. (Thomson 1974, p. 30)

The entire period between 1815 and 1871 was a period of tremendous political ferment across Europe, triggered by the French Revolution and the revolutionary wars. The defeat of Napoleon marked the victory of his conservative, reactionary, and anti-revolutionary enemies across Europe, including the brief restoration of royalty in France.

> As a result of the settlement at Vienna (1815), the ideas and institutions of hereditary, absolute monarchy were given a new lease of life throughout Europe. (Thomson 1974, p. 81)

This reactionary backlash across most of Europe, including Britain, lasted for a period of fifteen years from 1815 to 1830. Though much of the magic of the monarchies of Europe had gone, this backlash resulted in the restoration of the power of hereditary absolutist monarchies across Europe, the restoration of the privileges of the Roman Church which belonged to the forces of established order, and the restoration of the feudal system of land ownership. The link between land ownership and political power remained intact across Europe. There was a return to the evils of the Ancien Régime—patronage, corruption, and intimidation coupled with administrative inefficiency—fired anew, no doubt with a spirit of revenge. In Britain, the government suspended the Habeas Corpus Act, the bulwark of individual rights. It further passed laws designed to prevent large public meetings, undermine radical reforms, and to control the radical press.

In France, in 1815, a series of acts were passed by the Chamber giving the king the power to suspend individual liberties and the freedom of the press. In Austria (which extended across significant parts of Germany, Italy, and Eastern Europe), an infamous "system" was designed to keep public order through a network of spies, secret police, and terrorists. In every country, governments relied on informers, agents provocateurs, secret police, and military repression. In Spain, the republican constitution of 1812 was annulled and the prerogatives of absolutist monarchy resumed.

> Across Europe, rationalist ideas of the natural rights of man and secularist doctrines of State power were tarnished and even some of the greatest intellectuals of the times reaffirmed the dogmas of Christianity and old religious beliefs. (Thomson 1974, pp. 90 and 85)

While the French Revolution had held out the promise of universal suffrage—the right to vote, at least for the male part of the populace—the subsequent reactionary backlash across Europe sought to restrict that right.

In France,

> The vote for the Chamber of Deputies went only to citizens above 30 years of age who paid at least 300 francs a year in direct taxation. The electorate was

> thus only some 90,000 in a population of 30 million. To be a deputy, a man had to be over 40 years and pay 1000 francs a year in direct taxes. (Thomson 1974, pp. 86–87)

Moreover, the power of the deputies was shared with an upper chamber, the House of Peers, consisting of the higher aristocracy and clergy. This guaranteed that the whole policy was highly conservative.

In Britain, "the electorate was determined by an antiquated and complicated system of property quantifications which gave the vote to some 400,000 men and effective power to the landed gentry in the countryside and the large landowners and men of wealth in the towns" (Thomson 1974, pp. 87–88); in a population slightly greater than 10 million around 1815 (Hobsbawm 1968, p. 326).

As a result of the Napoleonic wars,

> the whole of Western Europe between the Pyrenees and the Baltic was infused with a strange mixture of general sympathy for the original ideals of the Revolution and an immediate hostility to the practices of the French. It was the perfect mixture for nourishing the seeds of nationalism. (Thomson 1974, p. 31)

Another result of the Napoleonic War across all of Europe was the unification of the Christian churches across Europe and Russia in offering a common opposition to republican and secular ideas. The three major branches of the Christian church across Europe, Roman Catholic, Protestant, and Orthodox, it must be remembered, had been at daggers drawn for centuries:

> Even Protestant powers, such as Britain and Prussia, were prepared to support the revival of papal power in Europe; and they were backed in this by the other great non-Catholic power, Russia. (Thomson 1974, p. 84)

These "seeds of nationalism," sown during the years of Napoleonic War and conquest, germinated across Europe in the fateful year 1848, capturing the imagination of the young Marx (and his friend and co-author Engels), finding reflection in the stirring prose of the *Communist Manifesto*. Marx was, of course, keen that the laboring underclasses of Europe utilize this historic opportunity—when the feudal powers were on the wane and the capitalists not sufficiently powerful and well organized—to grab power and establish their own republics, reflecting the revolutionary ideals of the French Republic. The year 1871 saw the emergence of two powerful nation-states—Germany and Italy—formed out of the expansion of the core states of Prussia and Piedmont respectively, bringing to a conclusion the long nationalist ferment across Western Europe. The year 1871 also saw the collapse of the Paris Commune, the last attempt of the working class rebels of Paris to challenge the combination of conservative and bourgeois—liberal forces who had ensconced themselves in power. Marx, now well on in years, saw the sun set on his aspirations of successful revolution across Europe—he had spent the years between 1848 and 1871 exhorting the continental revolutionaries, having been exiled from continental Europe.

Sometimes, the complex history of an entire region can be encapsulated in the fate of a small territory. Alsace-Lorraine was such a territory, with a mix of Franco-German population. It was annexed by France under Napoleon in 1806, to be re-possessed by Germany under the Prussian Bismarck in 1870–1871. To fast forward history a bit (that is, beyond Marx's lifetime), Alsace-Lorraine was reverted to France in 1918 following the Treaty of Versailles at the end of World War I, only to be retaken by a show of force by Hitler in 1936 in a stunning abrogation of the terms of the Treaty of Versailles. The fate of Alsace-Lorraine, over a period of a century and a half, showed which direction the winds of war were blowing in Western Europe. Not to be overlooked beneath the fog of war and the competing rhetoric of the belligerent states, Alsace had a well-developed textile industry while Lorraine had rich deposits of coal, the fuel of the early Industrial Revolution.

Notwithstanding the enormous importance of the French Revolution for the rest of Europe (and the world), there was another revolution that swept across Europe at the same period, namely, the Industrial Revolution (Hobsbawm 1962/1996, p. ix). Historians, for once, seem to be in agreement that the Industrial Revolution started in Britain around the fourth quarter of the eighteenth century, from which it spread rapidly to France, the Low Countries, and Western Germany, and still later to Northern Italy, Eastern Europe, and even further east to Czarist Russia. By the time Marx started observing and narrating events circa 1844, the Industrial Revolution was well advanced in Britain and was accelerating across Europe. This Industrial Revolution brought to prominence an upwardly mobile capitalist class of manufacturers, tradesmen, business entrepreneurs, and financiers whose worldview was at distinct variance with that of the Ancien Régimes of Europe. Whereas the wealth harnessed by the members of the Ancien Régime had depended on the ownership of land and agriculture, the wealth created by the vigorous class of capitalists depended on harnessing human labor. This latter form of wealth soon started challenging the older wealth, privileges, and rank based on hereditary and dynastic principles which formed the bedrock of the Ancien Régime. Liberalism, the ideology of this emergent class, stood in opposition to the suffocating and oppressive ideology of the Ancien Régime and its inefficient management of the affairs of state, particularly during the period of reactionary backlash and restoration of royal, feudal, and ecclesiastical privileges. Marx, living in England where the process of industrialization was most advanced, could clearly see it unfolding across Western Europe. Even more to the point, he could also see its ugly underbelly, the creation of a dispossessed proletarian class whose merciless exploitation formed the fount of wealth of the emergent capitalist class.

It is extremely difficult for an observer in the twenty-first century to even begin to grasp the deep churning process of immiserization that was unleashed across Britain in the seventeenth and eighteenth centuries as a precursor to the Industrial Revolution and the rise of the factory system of production. The broad outline of this ugly process, which enclosed the commons, broke the backbone of the British peasantry, displaced them from their lands and communities, and eliminated them as a class from British political life, is well described by Polanyi (Polanyi 1944/2001, pp. 37–43) and Moore (Moore 1966, pp. 9–28). It is probably for this reason that Harvey keeps referring to "dispossession" while describing the spirit of neo-liberal "reforms" unleashed upon the developing world toward the end of the twentieth century (Harvey 2007).

Polanyi describes the enclosures as a revolution of the rich against the poor in Britain; the starving, immiserized, and hapless population provided the grist for the "satanic mills" of the emergent capitalist class of factory owners and their allies. A similar wave swept across Western Europe in the early nineteenth century. The defeat of the Prussian armies at the battle of Jena in 1806, which almost resulted in the extinguishment of the Prussian state, produced such a shock in Prussia that it resulted in a dramatic reorganization of the army as well as the machinery of government. Hardenberg, one of the principal architects of the latter process, is reported to have written in 1807 to the King of Prussia: "We must do from above what the French have done from below" (Thomson 1974, p. 100). (We must presume, when an Elder Statesman writes to his King, that he is discussing revolution and not sexual preferences.)

This "revolution from above" sowed the seeds of European fascism which were to germinate later in the first half of the twentieth century; they were also the first indications of the rise of militarism in Europe in the reminder of the nineteenth century.

This "revolution from above" did achieve a specific purpose across most of Europe, between 1847 and 1880. In 1847, most of middle Europe (and Prussia) east of Britain, France, and the Low Countries still continued with serfdom in the land holding system. By 1880, serfdom had been legally abolished across this entire region (Hobsbawm 1975, pp. 368–69), no doubt under the active influence of the liberal ideology backed by the growing power of capitalism across Europe. This growing strength of the capitalist class, in uprooting the conservative Ancien Régime which had been deeply entrenched in power for centuries, can be attributed to the wealth and power garnered by industrialization. This is the reason why Marx, from 1875 up to his death in 1883, concentrated on producing his magnum opus, *Das Kapital.*

We have noted earlier that the periodization of Marx's writings—from 1844 to 1871 and 1875 to 1883—is not arbitrary. In the context of European history, it marks the watershed moment in 1871 when the major national consolidation of Germany and Italy were complete, the flames of "revolution from below" had largely been ruthlessly smothered by a combination of the conservative Ancien Régime and the liberal capitalist class and the Great Powers of Europe then turned their eyes from domestic concerns to the acquisition and consolidation of colonies and empires across the world, even as a vast concentration of financial capital was developed in the countries of Europe.

It may be noted that the United States largely kept away from any involvement in the messy transformational politics in Europe, through it had briefly supported Napoleon in war, mostly because of its quarrel with its earlier colonial master, Britain, and because its own memories of independence through revolution were still relatively fresh. Looking ahead for a brief moment, this behavior of the United States is in marked contrast to its armed support in attempting, in the twentieth century, to crush the Russian revolution and the Chinese revolution, as well as the emergence of post-revolutionary states in South America right into the twenty-first century, a posture it continues up to the present. After the Napoleonic upsurge in France had been suppressed in Europe, the United States protected its interest, with the active help of the British Navy, by espousing the Monroe Doctrine ("hands off South America"). This intervention weakened the hold of Spain and Portugal over their South American empire and paved the way, between 1822 and 1830, for the emergence of independent countries in South America: Argentina, Colombia, Mexico, Bolivia, Chile, Ecuador, Paraguay, Peru, Uruguay, and Venezuela. In this way, the French Revolution and its Napoleonic aftermath had repercussions very far away from Europe. Britain was the beneficiary through its increased trade with the former Spanish and Portuguese colonies.

For the purpose of this monograph, we use the term Industrial Revolution as distinct from the processes associated with industrialization, the prior establishment of the system of capitalism on a globally expanding scale and the process of colonization of distant lands. As Beckert has pertinently observed in his global history of cotton:

> As Spanish, Portuguese, French, English and Dutch powers captured huge territories in the Americas, they took away the continents' moveable wealth: gold and silver. It was indeed some of these stolen precious metal that had funded the purchase of cotton fabrics in India in the first place . . . paid African rulers to go on a hunt for labor, exchanging captives for the products of Indian weavers . . . more than 8 million slaves were transported from Africa to the Americas, first mostly by Spanish and Portuguese traders, to be joined in the seventeenth century by the British, French, Dutch, Danish and others. . . .

> Never before had the products of Indian weavers paid for slaves in Africa to work on the plantations in the Americas to produce agricultural commodities for European consumers. . . . Three moves—mperial expansion, expropriation and slavery—became central to the forging of a new global economic order and eventually the emergence of capitalism. . . . The beating heart of this new system was slavery. (Beckert 2014, pp. 35–37)

By the time the steam engine arrived on the historical stage, most of the above processes including the spread of the factory system, the emergence of a proletariat based on wage labor, the appropriation of the commons, and the increasing use of machinery for production were under way in continental Europe. The steam engine, based on fossil fuel (coal), gave an enormous boost to the expansion of the capitalist system and its eventual dominance as a global system. Marxian analysis of political economy which is inclusive of nature probably remains the best way of grasping the enormity of the above historical processes and the consequences that flow out of them up to the present day.

A deeper understanding of nineteenth-century industrialization—the Industrial Revolution—is called for because this revolution has in the following century fed into all three of the current crises. First, it has fed into the generalized arms crisis (the nuclear arms race is the spearhead of a more general process of military confrontation) by providing the material and logistic support (the muscle) which only an industrial society can provide. Second, it has fed into the environmental crisis due to the hugely increased throughput of materials and fossil energy—a quantitative increase in the pursuit of which lands have been mined to destruction, natural forests destroyed (and sometimes replaced by uniform plantations which are prone to fire hazard, pest outbreaks, and habitat loss for other species), fresh water sources depleted, damaged, diverted, or polluted on a global scale, oceans desertified or rendered dead in large parts and the atmosphere polluted by carbon dioxide to an extent that the heat balance of the earth has been altered. Third, it has fed into the climate crisis by pumping carbon dioxide into the atmosphere through the use of fossil energy. We shall have more to say later about all three feeder processes which have resulted in the present crisis. For the moment, we have to take cognizance of coal as the major source of energy that started off the Industrial Revolution in the nineteenth century. It was coal that

> transformed societies and shaped the fate of nations . . . launched empires and triggered wars . . . fuelled the Industrial Revolution in Britain, propelling the rise of a small rural Kingdom into the greatest commercial empire in the world. (Freese 2003, back cover)

The standard historical narrative regarding the start of the Industrial Revolution in England toward the latter half of the eighteenth century accords pride of place to a single invention—the steam engine. However, as noted by Freese (Freese 2003, pp. 58–64), the invention of the steam engine was an evolutionary process, starting from the Newcomen engine in 1712 to the Watts engine in 1776. Most importantly, the invention of such a device had become a dire necessity because of the huge and growing appetite for coal all over Britain, particularly London. Other environmental historians have also noted that the rising population and the diminishing forests of Britain had forced a shift to coal, the disappearance of forests being linked to the demand for charcoal from wood, used for steel making. Although coal had been in use in Britain for three centuries if not more, its output from the collieries had always been constrained by the accumulation of water in the underground coal pits. Many pits had to be abandoned; the larger underground mines needed as many as fifty to sixty horses and their attendants merely to dewater the mines, greatly adding to the costs and hindering the productivity of the mines. Newcomen's engine, followed by the Watts engine, running on coal, provided a way past this constraint on productivity. In particular, the Watts engine produced four times more work output as compared to heat input from burning coal in comparison with the Newcomen engine and was widely and immediately adopted in the collieries, paying for itself in terms of the saved costs of horses and attendants. This significantly increased the (labor) productivity of coal mining and reduced the cost of coal while significantly increasing mine output. There is an interesting positive feedback system brought into being due to the adoption of the steam engine in dewatering, since the amount of coal burned in the steam engine was far smaller than the increase in coal output that it enabled; thus, investing in the steam engine was hugely profitable. Watts (and Boulton) also produced a second prototype of the steam engine which was applied to the blast furnace used for iron and steel production. Here, the bellows used for blowing air into the blast furnaces used to be operated by human or water power; when replaced by the steam engine (coal was already available in the forges, mixed with the iron ore for smelting), the increased and continuous air supply increased the productivity of the blast furnaces, reducing the cost of production of iron and wrought iron, another positive feedback in operation. The further modification of the stationary steam engine into a mobile engine, the steam locomotive, enabled the haulage of large quantities of coal over long distances, using coal itself as the source of energy: in effect, the coal hauled itself, in the perceptive words of Ralph Waldo Emerson (Freese 2003, p. 10). The lowered cost of iron and steel made the laying of railway lines for coal haulage not merely feasible but hugely profitable, another positive feedback system that grew rapidly. It was, after all, competing with the haulage of coal by horse-drawn wagons on wet, treacherous, and rutted roads or horse-drawn barges along canals leading to the sea.

There is perhaps a lesson to be learned here by the profession of economics: if renewable energy has to be rapidly expanded in the twenty-first century, it must be applied in those sectors and processes where it will trigger such positive feedback mechanisms. In other words, in those production places where the difference between no electricity and even small quantities of electricity can make a huge difference in factor productivity.

To return to our narrative, the Industrial Revolution spread very rapidly across many sectors of production precisely because of some combination of the above factors. The vast production of iron and steel enabled the production of many forms of machinery (and armaments, for example, the musket) and the railways hugely expanded the reach toward distant markets, both for raw materials and finished goods. Over time, the steam engine was also applied to shipping, consolidating the capture and colonization of distant lands. The building of iron-clad ships powered by huge steam engines and carrying heavy cannons was a logical step that made Britain the dominant naval power of the period. Another new invention, the telegraph, provided the means for rapid transmission of information, both private and public, over huge distances. There is no comparison between the speed of information transmission by telegraph in competition with the then existing modes of communication—a man on horseback or a messenger traveling by horse-drawn coach. The telegraph system provided a kind of nervous system for the control not only of vast hinterlands but of vast colonial empires acquired by the Great Powers of Europe. Though Britain and France had already begun the process of colonial acquisition prior to the French Revolution and the Napoleonic War, the other powers were not lagging far behind.

Marx would have seen these core technologies during his lifetime, both in Britain and rapidly spreading across continental Europe. With the luxury of hindsight that the twenty-first century provides, it is also important to take note of the technologies that Marx, who died in 1883 without completing *Das Kapital*, would not have seen during his lifetime. Marx would not have seen grid electricity (it was first produced using hydropower and thermal power in 1882 and 1883). He would not have seen the internal combustion engine, the commercial prototypes of which were produced in 1886 (the Otto engine or the petrol engine) and 1897 (the Diesel engine). Although crude petroleum had been discovered in 1869 in Pennsylvania, its use was confined to household lighting purposes. The invention of the internal combustion engine, using refined petroleum products for combustion, provided a huge new market for petroleum products which helped extend the transportation reach to distant markets, inaugurating the empire of oil in the twentieth century. One of the waste by-products of petroleum refining, asphalt or tar, provided the material for constructing roads, furthering the reach of automobiles through the construction of motorable all-weather roads. Marx would not have seen heavier than air flight, first undertaken successfully in 1908. This heavier than air flight would not have been possible without the internal combustion engine to power it; the first airplanes, rapidly deployed in the First World War, added the third dimension to earlier forms of warfare which had been hitherto confined to the surface of the earth—land warfare and naval warfare. Marxist historians would do well to identify the strict chronology and interconnections of technology development in laying out the further course of the Industrial Revolution after Marx. Nevertheless, even based on the limited range of technologies that Marx was able to observe during his lifetime, his analysis was sufficiently penetrating to provide a basis for understanding the course of the twentieth century and beyond.

This necessarily brief outline of the Industrial Revolution as it unfolded during Marx's lifetime brings us to the second half of Marx's writings—his magnum opus, *Das Kapital*, the project of writing which absorbed his energies from 1875 up to his death in 1883. As indicated earlier, the year 1871 was a watershed in European history marked not only by the completion of nation formation in Germany and Italy but also the last gasp of any revolution from below in the failure of the Paris Commune. Across every European country, the bourgeoisie had successfully tightened its grip on power, challenging the monopoly of the Ancien Régimes and beating back any revolutionary upsurge. The power structure of Europe seemed to have evolved into a modus vivendi wherein the Ancien Régime, with its obsession with security, was left in control of both military affairs and home security (police powers) along with foreign affairs which included prying into and meddling into the internal affairs of other nation-states which were regarded as potential threats to security. On the other hand, the bourgeoisie retained with themselves the control of the economy and society which permitted them to reshape all its institutions in their own image and their own interests involving state budgets, taxes, and subsidies and many other "secular" changes. As Thomson has summarized it:

> the total effect was to revolutionize the whole meaning and function of government and politics. Instead of being concerned only with general matters of public order and national security, government had to act at the deepest levels of social and economic life. . . . The old dynastic conception of ruler and subjects had to be replaced by the conception of a state and its citizens. This notion of government and society as mutually interdependent and enmeshed, with a state emanating from the community it governs and the community demanding the constant service of its state, was the most revolutionary notion of modern history. . . . It was the common basis of all the greatest movements of nineteenth century Europe: on the one hand nationalism, and on the other liberalism, democracy and socialism. (Thomson 1974, p. 98)

It may be inferred from the above modus vivendi between the two dominant classes that so long as substantial budgetary allocations are made for defense, home security, and foreign affairs which are the domain of the conservatives, the bourgeoisie is free to use the balance of the machinery of state for its own interests. This also appears to explain why these three budgetary allocations are treated, in modern western societies, as if they are sacrosanct; they represent the guaranteed "cut" to the conservative powers within society. After the various elements of the bourgeoisie are through with appropriating their share of the societal budget—through subsidies, tax concessions, rebates, supportive measures, and duty drawbacks (a whole veritable lexicon has been invented to mask this diversion of public funding), the leftover balance is then applied to the rest of society, which is the vast majority of citizens. There is an unintended irony in the phrase "trickle-down theory" as applied to the workings of the national political economy because that which reaches the bottom of society—the poorest sections—is but a trickle. This is the heart of the explanation of the huge disparities of wealth and material possession that is evident in developed capitalist economies; and bourgeois economists who suddenly "discover" these disparities (say in the latter half of the twentieth century) need to be commended on their historical myopia.

To return to Marx and *Das Kapital*, Marx realized that a full-fledged analysis of Capital (with a capital C) was necessary to understand the new order of affairs that had emerged. Marx had begun this process in earlier writings, notably "Economic and Philosophic Manuscripts" (1844) and "Grundrisse—Foundations of a Critique of Political Economy" (1867). *Das Capital* is a mature work of political economy—there is no artificial separation of economics from politics and the further mystification of economics as a "science." The treatment is of real men (and women, children) involved in real processes of production, exchange, and consumption and the real historical processes that have made them what they are—this is what Marx deems to be "scientific enquiry" because it unmasks surface appearances and constantly probes for the underlying networks of causes, motives, and class interests. Throughout, the treatment is international in scope, not falling into the trap of narrowly competing nationalisms, because nationalism is itself a manufactured, fabricated construct sub-serving class interests and masking them.

We shall not be summarizing *Das Kapital* here—that task has been ably performed by David Harvey in a series of thirteen lectures (available on YouTube) on Volume 1 of Capital and nine lectures on Volumes 2 and 3 of Capital (Harvey 2008 onward). Harvey has also written a guide to the lectures on Capital (Harvey 2010) which provides an overview of the substantive arguments in the three volumes of Marx's *Capital*. Only the first volume of *Das Kapital* was published in German during Marx's lifetime. Volumes 2 and 3 were compiled by Engels from the voluminous notes left behind by Marx at his death.

In his thorough, historicizing, Germanic, and critical way, Marx has looked not only at the anatomy of capital as a system of the production, exchange, and circulation of commodities but also their relationship to labor, money, and the various classes in society—their emergence and role in the production and reproduction of capital. To appreciate Marx's contribution fully is to acknowledge his contribution to philosophy and diverse fields of the social sciences: political economy and politics, history, law, sociology, anthropology, and geography as well as literature and in general, critical studies in the social sciences. Whereas the thrust within academia and the universities has been on greater disciplinary specialization, Marx challenges the reader by ranging across fields and disciplines, which makes the reading of Marx such a challenging endeavor. Marx, it must be remembered, is not writing for an academic audience but he is constantly self-aware that his writings will be subjected to vicious ideological attacks from the ruling classes. Marx did predict that eventually capitalism would collapse because of its internal contradictions—the contradictions between the capitalist class and its predecessor classes of the Ancien Régime, the contradictions between the various sections of the capitalist class (the financial capitalists, the industrial capitalists, and the commercial capitalists, all vying to increase their appropriated share of the social surplus), and most importantly the contradiction between the capitalist class and the proletarian class, the decisive one that would sweep away the ruling classes.

At the risk of being misunderstood, one would say that Marx stands in relation to the social sciences in the way that historically Newton stood to the natural sciences, in particular physics and mathematics (we are including mathematics with the natural sciences because of the close co-evolution of the two historically in the early modern period: purists may kindly overlook the transgression!). It is undeniable that Newton's contributions toward the physics of his time ranged from the theory of gravity to optics, mechanics, dynamics and kinematics, fluid dynamics, and thermodynamics; in mathematics to the invention of calculus (along with Leibnitz); we have already noted the impact of Marx's writings across several fields of the social sciences. The chief difference is that Newton had the full backing of the establishment of his time—he was a Fellow of the Royal Society and its recognized leading light, whereas Marx had the full opposition of the establishment of his time to contend with. A full appreciation of Marx needs to record that he worked under the most trying of circumstances—an international political refugee, financially supported by his friend Engels, constantly having to defend himself in the courts of law against various attacks on his writings and political activities—to leave behind a corpus of writings for future generations. Against all these odds, he was the public intellectual of his times, the fountainhead of so many of the streams within social science and philosophy. We have to stress both aspects—the "public" and the "intellectual"—because, as subsequent history has shown, regimes that are fascist or authoritarian invariably clamp down on publicly voiced dissent and are anti-intellectual in their ideological orientations. They are threatened by intellectuals who can unmask the shabby hypocrisy of a "revolution from above."

We are deliberately omitting Marx's influence on socialist revolutions in the twentieth century—there is a tendency to blame him for every action and consequence of later events, a logical fallacy which states that "if event Y follows event X then event Y must have been caused by event X"; this logical fallacy is now referred to in shortened form as the post hoc fallacy and a surprising number of establishmentarian reputations have been built on exploiting this fallacy, to the benefit of the establishment and the denigration of Marx.

At least on the occasion of two hundred years of his birth, let there be a mark of grateful remembrance and gracious acknowledgment of this public intellectual. The best such acknowledgment by later generations would be to read what he has written, grapple with the difficulty of the subject matter, and forgive him his trespasses—not that it makes any difference to him now but who knows? It might make a difference to us and our future.

Chapter Three

Marx and the Twentieth Century

Before bidding adieu to Marx and the nineteenth century and moving the narrative on to the twentieth and then the twenty-first century and the main themes of this monograph, three themes have to be flagged for further discussion:

1. the complex political relationship between class war, civil war, and war between nation-states and blocs of nation-states
2. the relationship between the Marxian analysis of capitalism and the Industrial Revolution and its relationship with nature
3. the relationship between falling demand, economic deflation, and the collapse of capitalism

Each of these is related to the crises confronting the twenty-first century, in specific ways. We shall deal with the inter-relationships in a later section.

INTERREGNUM: 1875–1914

The period from 1875 to 1914, referred to by Hobsbawm as the Age of Empire (Hobsbawm 1994), was an extremely important period during which the Great Powers of post-Napoleonic Europe consolidated their worldwide colonies and expanded their empires. Even during the course of the Napoleonic wars, Britain and France had clashed in America and Canada in the second half of the eighteenth century in India and Africa. Toward the end of the nineteenth century, the other powers also joined the race for foreign colonies, in which Britain and France had the lead. Thomson shows the extent of colonial acquisitions on the African continent (Thomson 1974, p. 463): there is a simple pattern to it. If a line is drawn from Cairo in the north to Cape Town in the south, the British holdings were to the east of the line in a more or less continuous tract from south to north, ranging from (present day) South Africa, Botswana, Zimbabwe (Rhodesia), Kenya, Uganda, Sudan, and Egypt. This was referred to as Anglophone Africa. The French holdings mostly lay to the west of the same line covering the (present day) countries Tunisia, Algeria, Morocco, most of the Sahara desert, Senegal, Gambia, Ivory Coast, and part of Congo, referred to as Francophone Africa. Since both were interested in having landing points for their ships sailing toward India and further east, the British had strategic points on the West coast of Africa in Sierra Leone, Gold Coast, and Nigeria. Likewise, the French had landing points in the French Congo and the island of Madagascar off the east coast. The Germans had territories in Cameroon, Namibia, and Tanzania, two being on the west coast and one on the east coast. The Belgians had colonized the huge and rich territory of Congo. The Portuguese had a huge territory in Angola on the west coast and Mozambique on the east coast. The Italians had Libya in North Africa and Somalia and Eritrea on the east coast. However, the territories colonized by Britain and France, taken together, were far greater than those of all the other continental European powers taken together. This was a tinder box waiting to explode.

Further to the east, the pattern of dominance in colonial holdings were similar: Britain holding major parts of (present day) Pakistan, India, Bangladesh, and Burma as also Australia, New Zealand, and Singapore, the French holding (present day) Vietnam and Cambodia, the Portuguese with small holdings in Goa (India), the Malaca Straits, and Macao, the Dutch with holdings in Indonesia. The British also held several islands across the Atlantic to the west of Africa, the West Indies, which were profitably operated using African slave labor for supplying sugar to Britain. These colonial possessions provided resources by way of raw materials for industrialization (for example, cotton), provided captive markets for industrial products, provided outlets for foreign capital investment, and cheap labor on demand. Indeed, the social peace in Europe was obtained, along with the suppression of the politics of the rebellious masses, by liberal doses of surplus extracted from the worldwide colonies, which also absorbed migrants from home countries therewith relieving part of the population pressure in Europe. In a half century or less, Europe was on the path to prosperity while its worldwide colonies had their heads pushed under the water. Perhaps a single illustration, cited by the eminent historian Fernand Braudel, may serve to explain one of the underlying mechanisms at work:

> "The labour of a foreign proletariat" was the very foundation of Europe's trade with India . . . defending the principle of bullion exports in 1684, Thomas Mun [argued]: the Indian goods which the East India Company had bought at a cost

> of £840,000 had been sold in Europe for 4 million—so they effectively brought money into England. (Braudel 1984, pp. 521–22)

Multiply such transactions across several centuries and across the hapless colonies of the world and we begin to get a dim idea of how European prosperity came about at the cost of the rest of the world, as also the intensity of European class interests in preserving and expanding this relationship with the rest of the world. It deserves to be noted that the United States, up to 1918, did not participate in the worldwide land grab that the European imperial powers and Japan indulged in. The United States had enormous land resources for expanding its agriculture and food supply, enormous coal (its coalfields, the richest in the world, covered a land area equal to half the area of Europe [Freese 2003, p. 105]), and other minerals to power its industrialization and an abundance of labor power which could be expanded; ergo, it did not need colonies. The other states that did not participate in worldwide colonization were the Scandinavian states.

While Europe enjoyed a century of relative peace (from the end of the Napoleonic wars in 1815 up to the start of World War I in 1914), the wars between the European powers were mostly fought over the colonies outside of Europe. Yet, within Europe itself, the Austro-Hungarian Habsburg Empire, the largest territorial state in Europe, was clearly tottering. It had lost vast chunks of territory to Prussia and Italy in 1871. It had lost its colonial possessions in Latin America. Its industrialization had none of the vigor of its surrounding European powers—Britain, France, and Germany. Neither did the Turkish Ottoman Empire, which had borders in Eastern Europe touching the Austro-Hungarian boundaries, have the level of industrialization to forcefully withstand the power of the surrounding empires in Europe and elsewhere. Covetous eyes were turned on the territorial possessions of these weak empires.

This was the tinder box that caught fire in 1914, not only in Europe but worldwide due to the colonial spread across the world.

THE SHORT TWENTIETH CENTURY: 1914–1991

Hobsbawm has used the above title to compress the major part of the twentieth century's major events into the above period (Hobsbawm 1994): respectively World War I (1914–1918), World War II (1939–1945), and the Cold War (1945–1990). Hobsbawm has fused the two world wars into a single "Age of Catastrophe," particularly for European civilization but with enormous consequences for the rest of the world. Historians have tended to view the nineteenth century as the "age of bourgeois civilization," followed by the catastrophic twentieth century. However, as Beckert pointedly notes about both centuries,

> Looked at from the perspective of much of Asia, Africa and the Americas, one can argue just the opposite—that the nineteenth century was an age of barbarity and catastrophe, as slavery and imperialism devastated first one pocket of the globe and then another. It is the twentieth century, by contrast, that saw the weakening of imperial powers and thus allowed more of the World's people to determine their own future and shake off the shackles of colonial domination. Without its Eurocentric distortions, decolonization would be at the very centre of the narrative we tell about the twentieth century—and this retelling would allow us to see that global capitalism today is most fundamentally shaped by the struggles for independence. (Beckert 2014, p. 441)

We intend to follow the more conventional view of regarding the two world wars as separate but linked events which resulted in the emergence of the two major superpowers—United States and (the then) USSR—and the prolonged confrontation between them, referred to as the Cold War.

In very brief outline, of the seven imperial powers jostling each other in continental Europe, four collapsed in the aftermath of World War I: the Austro-Hungarian Habsburg, the Turkish Ottoman, the Russian Tsarist (the latter succumbing to revolution), and the German Second Reich. At the end of World War I, Germany lost her colonies in Africa, the territories being taken over by Britain and France in accordance with a secret agreement between them prior to the Great War. Germany also lost Alsace-Lorraine and its coal as well as part of its Polish territories. One of the consequences of the First World War was that "no old government was left standing between the borders of France and the Sea of Japan" (Hobsbawn 1994, p. 29).

None of the defeated empires escaped revolution. The main claimants to the succession (of the collapsed empires) were various nationalist movements (Hobsbawm 1994, p. 31). Among the other consequences of the First World War was its impact even on the victorious powers: Italy was wracked by internal turmoil, Britain and France driven into physical exhaustion and bankruptcy.

As far as Europe was concerned, the First World War was a mass war with massive mobilization of resources, both material and human, by the contending imperial powers. For the first time in history, the powers of industry applied to killing enemies was on full display and the results were tragic in the trenches. Men, often mere boys, numbered in millions among the dead, those not killed outright on the battlefields often succumbing to infections and gangrene away from the trenches. An entire generation was traumatized, brutalized, and decimated by the war and all of Europe sickened in its aftermath. Yet the slaughter might have proved indecisive, between eventual victors and vanquished, had not the United States intervened in 1917 on behalf of the Allied powers: Britain, France, and Italy.

When we say "the powers of industry applied to killing enemies," it devolves upon us to take a quick survey of the range of new technologies and devices with which the First World War was waged. Transportation to and from the battlefield had been motorized with the invention of the internal combustion engine. Troops and supplies were hauled over long distances by the railways. The chemical industry which produced fuel for the engines also produced more powerful explosives for shells and bombs and poison gas. Warships were driven by steam power, the earliest aircraft used for war purposes were engine driven. Submarines were already operationalized and were a significant factor in naval warfare. Above all, the machine gun on the battlefield perhaps captures most vividly the image of the war, with its leather belt like a conveyer belt carrying shells as an input to a "machine" gun, whose output was a steady stream of bullets directed at the bodies of a charging enemy. War became three dimensional, and the range of technologies multiplied the force applied on the enemy much beyond any hand-to-hand warfare. Sheer quantities of materials produced and delivered to the "front" started mattering though the rhetoric of "bravery on the battlefront" remained unchanged. Needless to add, in the economic sphere, war demand could not be questioned, war supplies could be requisitioned through expanded powers of government; even able-bodied men could be requisitioned, hastily trained, and rushed to the battlefronts; and there was money to be made in providing war supplies, since prices could hardly be questioned. Money, moreover, was also made via war reparations imposed on a defeated Germany at the end of the war, as recorded and protested against by Maynard Keynes in his book "The Economic Consequences of the Peace." The reparations were bills for damages incurred by private parties, often hugely inflated, taking full advantage of the golden opportunity so provided. When cash or gold was no longer available with the Germans, any other material commodity was impounded and sold, often at a fraction of its value, to recover the damages. Keynes, then a young mandarin in the economics ministry of Britain, had the foresight to warn against this short-sighted greed; sure enough, it broke the backbone of the German economy, yet Keynes' voice was a cry in the wilderness. This experience was probably decisive in Keynes' later formulations in economic theory.

We have already noted earlier Hobsbawm's observation that no old government was left standing between the borders of France and the Sea of Japan and that none of the defeated empires escaped revolution. With their economies shattered and governments bankrupted all across Europe, the stage was set for the rise of fascism across Europe: Hitler and the Nazi Party in Germany, Mussolini the originator of the ideology of fascism in Italy, Generalissimo Francisco Franco in Spain, Antonio Salazar in Portugal, Admiral Horthy in Hungary, and later General Petain in France. The fear of internal revolution coupled with the presence of a socialist revolutionary worker's state in the form of the USSR undoubtedly acted as a spur to the political establishments across Europe just as they had done a century earlier with Napoleon and the revolutionary republic in France. Note that most of the European dictators came with a military background: Hitler had been a corporal in the Austrian army, which apparently gave him the military competence to overcome senior generals in the German army regarding critical decisions in the course of the Second World War. Hitler made the fateful mistake of opening a second front against the USSR, the same mistake that Napoleon had made a century earlier. As in Napoleon's case, the opening of a second front split his military forces into two, which was to prove suicidal in the event. His underestimation of the military and other resources of the USSR as well as the rigors of fighting a war through the bitterly cold Russian winter sealed the fate of his regime and that of the German people. It also sealed the fate of the other fascist dictatorships across Europe, at least those which had sided with him. Franco and Salazar had remained neutral in the Second World War; their regimes survived for a decade or two after the end of the war, to be overthrown by popular internal revolt.

The parallelism between the key events of the history of the nineteenth century and the history of the twentieth century across Europe is the major reason why we have tried to describe the nineteenth century in some detail, in the hope that it makes the twentieth century more comprehensible to the current generation. It is also our hope and endeavor that the history of both these centuries, taken together, will shed light on the twenty-first century.

The period between the two wars (1918–1939), which saw the rise of fascism across Europe, was also a period of worldwide economic depression across the capitalist system, a depression so severe the likes of which had never been seen before. This depression, triggered by a rapid collapse of demand, led to a closure of businesses which led to widespread unemployment further feeding into falling demand and deflation—another self-reinforcing feedback process—no doubt further exacerbated the fears of the ruling classes. It was a near classic enactment of Marx's analysis of the collapse of the capitalist system which kept the working classes on meager wages thereby depressing demand and leading to repeated and prolonged economic and political crises. Fascism in Europe was one response to suppress an imminent social uprising, by trying to turn the unemployed against the employed and fanning the insecurities of the lower middle classes who feared that they would slip into the ranks of the working classes (Moore 1966, pp. 433–42; Moore 1978, pp. 420–23; Eatwell 1996). The other response was war.

As indicated earlier, of the seven contending imperial powers in Europe, the First World War had resulted in the destruction of four, the Austro-Hungarian, the Turkish Ottoman, the Czarist Russian, and the German, with victory to the remaining three, Britain, France, and Italy. It nevertheless left behind a highly unstable, economically shattered, and politically resentful continent. Fascist Italy switched allegiance to a rising and increasingly powerful Germany under Hitler, tilting the earlier balance of power so that it became Britain and France versus Germany and Italy, that is, two versus two. Not that Italy was much of an industrial, military, or imperial power to contend with but its shift of allegiance provided the fig leaf of legitimacy that Germany under Hitler needed, to question and destroy the post–World War I order that was sought to be created particularly by the late intervention of the United States.

The Second World War destroyed the remaining European imperial powers, Britain, France, and Italy, as well as the aspiring imperial powers Germany and Japan, the latter having thrown in its lot with Germany and Italy. It also rained vast destruction on the USSR, the successor state of Czarist Russia, which had been drawn into the war despite initially having no imperial ambitions, due to the aggression from Germany. It needs to be recorded, at least for later generations, that the USSR had tried from September 1938 to August 1939 to engage with Britain and France to enter into a military alliance against the designs of Nazi Germany, only to be thwarted and rebuffed repeatedly by the naïve and incompetent British prime minister, Neville Chamberlain. Such an alliance might have stopped Hitler in his tracks through a clear early signal that he would have had to fight a war on two fronts—a Western front against France and Britain and an Eastern front against the USSR. With no such military alliance in place, the USSR signed its own treaty of non-aggression with Germany, to protect its own interests in peace. Indeed, following the Soviet revolution, the subsequent war with the White Armies, and the internal purges under Stalin (which included the virtual decapitation of the army and imprisonment of senior army leaders whom Stalin feared), the USSR was in no position to fight another war. The Russo-German treaty was to last for all of two years before Hitler violated it, as he did other treaties (Shirer 1950, pp. 569–77, 638–46, 655–64, 668–77, 700–705, 1062].

The two turning points of the Second World War, commonly regarded by historians, are the Battle of Stalingrad and the Battle of El Alamein. Interestingly, both have to do with the capture of oilfields respectively under Russian and British control. Part of the strategic aims of the war and the underlying politics become clear from the following quotes:

> "If I do not get the oil of Maikop and Grozny" Hitler told General Paulus, commander of the ill-fated Sixth Army, just before the summer offensive began, 'then I must end this war.' Stalin could have said almost the same thing. He too had to have the oil of the Caucasus to stay in the war. That was where the significance of Stalingrad came in. German possession of it would block the last main route . . . over which the oil, as long as Russians held the wells, could reach central Russia." (Shirer 1950, p. 1188)

> It seemed to many a startled Allied Statesman, poring over a map, that nothing could now prevent Rommel from delivering a fatal blow to the British by conquering Egypt and then, if he was reinforced, sweeping on north-east to capture the great oilfields of the Middle East and then to the Caucasus to meet the German armies in Russia. (Shirer 1950, p. 1191)

For those who will see, there is a clear line to be drawn between the First World War and the takeover of the Arabian oilfields from the Ottomans by the British and the Second World War and the defense of those oilfields from German occupation. This line can be further extended into the second half of the twentieth century and the beginnings of the twenty-first century. The Russian revolution had taken away the Caspian Sea oilfields from capitalist control but the Arabian oilfields represented the largest reserves of oil in the world. Ghawar oilfield in Saudi Arabia alone covers 2500 square kilometers of oil; according to oil geologists, there will never be another Ghawar in world history.

Earlier, we briefly indicated the range of new technologies and devices that were developed and applied in the course of the First World War. Applying the same procedure to the Second World War enables us to appreciate the massive killing power developed and amplified through a variety of technologies. The First World War was largely fueled by coal, since oil and the internal combustion engine had but recently appeared on the scene. The Second World War was largely fueled by oil: tanks and trucks to carry troops and supplies, more powerful aircraft to carry heavier payloads of bombs, the bombs themselves having more explosive capabilities using petroleum derivatives, ships and submarines using internal combustion engines and electric batteries, jet engines using turbines, missiles pioneered by the Germans and the technology later acquired by the United States in the person of Wernher von Braun, radar pioneered by the British, atomic devices based on enriched uranium and plutonium (hence two bombs had to be dropped on Hiroshima and Nagasaki respectively to test each technology) and the computer developed for crypt-analysis (code breaking) through the combined efforts of two of the leading mathematicians of the twentieth century—Alan Turing and John Von Neumann. By the end of the Second World War, war had become four dimensional: the three spatial dimensions of the First World War now reinforced with the addition of a fourth dimension, computerized cryptography, which was to be greatly expanded along with other applications of the computer to the paraphernalia of war in later years.

This listing is merely indicative but one can see both a historical evolution and a genealogy of technologies between the nineteenth and the twentieth centuries. The early steam engine was an external combustion engine in which the external boiler was used to generate steam which was then supplied to an arrangement of pistons, cylinders, and cranks to convert linear into rotary motion. This was followed by the internal combustion engine where the fuel was burned inside the cylinder to generate linear motion converted by a crank mechanism into rotary motion. The steam engine evolved into the steam turbine which eliminated the troublesome and speed-limiting crank, directly yielding rotary motion, higher speeds, and greater power output. Likewise, the i-c engine evolved into the liquid fuel/gas turbine, similarly yielding much higher speeds and power output. The invention of grid electricity based mostly on coal and steam further freed factory production from locational constraints thereby enormously enhancing output in every field of production. The by-product of petroleum production was a range of plastics, solvents, and organic compounds which have endeared themselves to the environmentalists, as also a waste product called asphalt used for making roads for running automobiles—the throughput of each of these in the world economy was enormously enhanced by the Second World War.

Many of the technological inventions were paid for through public funding; they were later to generate enormous profits for private sector corporations. Computers, civil aviation, and power generation are easily identifiable technologies and indicate why access to state power has been so important for generating huge private fortunes both during and after the war.

World War II inaugurated "total war" in which civilian populations were also targeted by both sides, particularly using aerial bombardment (Hobsbawm 1994, p. 43). At the end of the war, only two powers were left standing on the global battlefield: the United States and the USSR. Empires and the age of empires was truly over, to make way for an age of super powers amid a sea of newly independent nation-states, stretching across the globe.

The collapse of the empires liberated the nations of Asia and Africa from the colonial yoke. If the empires at the end of the First World War had been physically exhausted and bankrupt, at the end of the Second World War they were physically shattered and in a state of economic collapse, such being the price of war once the trumpets of martial glory have receded into silence. Most of the countries of Eastern Africa and South Asia were rid of the British as those of West Africa were rid of the French. Vietnam was among the exceptions and resulted in a war of liberation against the French, supported now by the United States. The British held on to the Falklands/Malvinas off the coast of Argentina, resulting in a war with Argentina in the eighties. In most other countries, the successor regimes were national-democratic states. China had undergone a peasant-based revolution and emerged as a second socialist republic alongside the USSR on the world stage.

1945–1991: THE COLD WAR AND THE GOLDEN AGE OF CAPITALISM

The two super powers—the United States and the USSR—had been allies in the world war to crush the fascist powers of Europe, yet a lasting peace between them was not to be. In rapid order following the war, the two power blocs confronted each other: the North Atlantic Treaty Organization joining the United States and the Western European nations and the Warsaw Pact uniting the USSR and the satellite countries of Eastern Europe occupied militarily after the Second World War.

US capital helped reconstruct Western Europe after the war; the US economy had, during the decade around the war, clocked annual rates of growth of around 10 percent per annum, putting an end not only to its domestic economic depression via a New Deal but to being strong enough to help Europe after the war. On the other hand, the Soviet economy was not capable of any such revival in Eastern Europe, having itself been badly destroyed further as a consequence of war. Nevertheless, both the Soviet Union and the satellite states in Eastern Europe registered high rates of economic growth following the end of the war (Hobsbawm 1994, p. 259). The collapse of empires resulted in a power vacuum and geo-politics, like nature, abhors a vacuum. This set the stage for the Cold War, so named not because guns and bombs were used by the two super powers against each other but rather that the age "lived and thought in terms of World War." This confrontation was to last up to the collapse of the USSR in 1990, which sent shock waves around the world.

The first half of the Cold War period (from 1950 to 1975) was also a period of sustained economic growth; it is often referred to as the Golden Age of Capitalism. According to Keynes' theory which enjoined upon the state to intervene and restore falling demand in the economy by re-routing the economic surplus from the top to the bottom of society, this intervention was necessary initially to restart economic circulation which had ground to a halt (Keynes' famous expression "digging holes in the ground" captures the essence of why incomes had to be delivered to the bottom of society to restore mass employment, wage-based incomes, and demand for goods). In sum, Keynes recommended the restoration of the system of production to "full employment, full output" through state intervention. This was the equivalent of heresy to those who believed in the ideology of "free markets" with a theological fervor. This confrontation between the heretics and the true believers continues in all the capitalist countries until date notwithstanding ample evidence that markets have performed well when bolstered by supportive legislation, hidden and open subsidies, and incentives and regulation by state institutions—in other words, markets have performed when they have not been "free" and precipitated economic crises when allowed to be free.

Across the Iron Curtain, a similar result was sought to be achieved through state-centric five year plans. Essentially, this was a period of reconstruction of economies ravaged by the war; and countries which had populations that were educated and industrially skilled were able to restore themselves with the provision of capital (tools, machinery, infrastructure, etc.). The capitalist countries also implemented welfare measures across the board in most cases: publicly subsidized health and education, doles for the unemployed paid for by the state through taxation of the employed, and so on. Fortunately, during the period, international oil prices were also low, averaging USD 2 per barrel for Saudi Arabian oil in the entire period from 1950 to 1973. With the formation of the oil cartel OPEC, "oil prices more or less quadrupled in 1973 and more or less trebled again at the end of the 1970s" (Hobsbawm 1994, p. 262 and p. 473). This roughly twelve-fold increase of oil prices in about six years torpedoed the Golden Age of Capitalism but also led to a prolonged cycle of energy shortages, increasing debt, collapse of global primary commodities prices as Third World countries fell over each other to export whatever they could at whatever price they could, to pay for expensive oil imports. The oil crisis of the seventies threatened the entire process of industrialization and modernization almost across the entire Third World barring a small number of countries which had their own oil. By 1980, oil and its technologies had become for the world economy what coal and its technologies had been in the eighteenth and nineteenth centuries for Western Europe and the United States: the principal source of energy for the transportation and circulation of all commoditics, sometimes even for electric power generation. This central feature of the political economy of the late twentieth century is often lost sight of in the over-profusion of economic "explanations" that have been churned out by often ill-informed observers to explain the events of the twentieth century. In brief, any country which has oil can power its own development and consequent prosperity autonomously (an oil substitute, technologically produced by the Fischer-Tropsch process as Germany did during WWII, can compete with oil at about USD 54 per barrel of oil; South Africa has followed this route, which is why its economy remained strong despite political vicissitudes). Countries without oil are rudderless in the international economy. Since Europe is largely dependent on imported oil and its price is set internationally, any drastic increase in global oil prices reduces economic growth across the EU, the largest economic bloc in terms of GDP in the world, and this has ramifications across the entire global economy. Venezuela, which has the largest reserves of oil in the world after Saudi Arabia, is now the focus of a geo-political struggle over the control of oil; its reserves, estimated at nine hundred billion barrels of oil, single handedly equal about thirty years of world oil consumption which is at around thirty billion barrels per year. This dependence of the major part of the global economy on oil has huge implications for climate change in the twenty-first century, to which we shall revert later in this monograph.

The oil price shock brought an abrupt end to the Keynesian welfare state politics in the capitalist countries, as economic conservatives came back to power. On the whole, Keynes had had a humane and internationalist outlook underlying the restoration of the international economic system back to working order. Part of this economic prescription had included the establishment of three international institutions to prevent the re-emergence of the kind of economic crises that had precipitated economic collapse and war: the International Bank for Reconstruction and Development (the IBRD, better known as the World Bank), the International Monetary Fund (IMF), and the International Trade Organization (ITO), later resurrected as the World Trade Organization. Shorn of their theoretical niceties, these institutions were to be designed respectively to provide investment capital, working capital, and to oversee rule-bound commodity trade at the level of national economies, particularly to the newly independent countries of the Third World.

Yet, the Keynesian prescription had a darker side to it, possibly unintended, that of military Keynesianism, that is, that demand and unemployment in the economy also could be resurrected through state-directed investment into the production of military goods, military infrastructure, and expansion of the military bureaucracy in all the four dimensions mentioned above. After all, employment in military production or the military bureaucracy also reduced the pool of unemployed labor, generated incomes to households, and generated the demand for material supplies from the larger civilian economy (for example, the Pentagon was reputed to be the largest single buyer of petroleum in the already vast US economy). This saw the emergence of the "military-industrial complex" (Hobsbawm 1994, p. 236) in the United States, mirrored in other Allied capitalist economies and mirrored likewise in the opposite camp, the centrally planned socialist economies.

This was the economic mechanism underlying the Cold War (or the arms race) and shared nothing with the originally humane Keynesian perspective. In fact, Hitler had already succeeded in Germany in putting the economy back on the rails by boosting military demand for steel to build warships, tanks, guns, planes, etc. even before Keynes had propounded his theory. There is a lesson to be learned here, perhaps, that humane intentions do not necessarily lead to humane consequences, as they did not at both national levels and at international levels. The working of the three Keynesian international institutions were duly misused to further the economic and political objectives of the capitalist powers.

The Cold War has left its mark on an entire generation if not more in both the capitalist world and the socialist world; not only has it created a stockpile of weapons both nuclear and conventional which is in consonance with the unarticulated doctrine of "total war." Its shadow has been cast over everyday life lived under the threat of mutually assured destruction (MAD), about the only restraint on direct confrontation between super powers. The world has turned its back on chemical and biological weapons which are barred under the Geneva Convention. It has not yet turned its back on nuclear weapons as weapons of mass destruction (WMDs) as though killing humans in extremely large numbers using nuclear weapons is somehow ethically more justifiable than killing them in large numbers using chemical and biological means. This is the reigning hypocritical madness that as yet governs the world, yet questioning it runs the risk not merely of being shouted down but of having the most obnoxious of motives attributed to those who dare to question.

Nevertheless, even as we have written the above words, today's newspaper carries the following short news item, incidentally datelined Washington, which we are reproducing in full:

> Headline: Agent Orange still polluting environment: Washington; Agent Orange, a toxic herbicide that the US aircraft sprayed during the Vietnam war, continues to enter the Vietnamese food supply 50 years later, a study has found. During the Vietnam War, US aircraft sprayed more than 20 million gallons of herbicides, including dioxin contaminated Agent Orange, on the country's rain forests, wetlands and croplands. Agent Orange defoliated the thick jungle vegetation concealing the Viet Cong fighters and destroyed a portion of the country's food crop. (Indian Express, 2 March 2019, p. 15)

If this is not an instance of chemical warfare then we probably have misunderstood the meaning of both terms "chemical" and "warfare." Moreover, almost all conventional high explosives used in modern warfare originate in the petroleum refinery so that almost all warfare is chemical warfare. More recent examples could also be cited: the use of the Mother of All Bombs (MOAB) in Afghanistan, the lethal device being a fuel-air mixture bomb with detonation energy almost equivalent to a tactical nuclear device, or the use of depleted uranium in armor piercing shells, etc. If "we" use such chemicals against "them" then we are justified because we have self-certified that we are moral. One can only shake one's head in sadness and quote Lewis Carroll:

> Said the mouse to the cur,
> "Such a trial, dear sir,
> With no jury or judge,
> Would be wasting our breath,"
> "I'll be judge, I'll be jury,"
> Said cunning old Fury;
> "I'll try the whole cause,
> And condemn you to death."
> (Carroll, *Alice's Adventures in Wonderland*, 1865)

To return to the real world of politics and history, the Keynesian prescription, applied across the capitalist world but particularly to the United States and Western Europe, had another consequence: world output of manufactures quadrupled between the early 1950s and early 1970s; agricultural output also shot up and world fisheries trebled their catches (Hobsbawm 1994, p. 261). Since the socialist part of the world economy also grew as did economic output from most of the Third World countries, there was a resultant impact on the global environment. As Hobsbawm (p. 270) points out, much of the new industrialization of these decades was the spread of old industrializations based on old technologies to new countries—the nineteenth-century industrialization of coal, iron, and steel to the socialist agrarian countries, the twentieth-century American industries of oil and internal combustion engines to European ones. Yet again, it was based on an expansion of fossil fuels and attendant technologies. However, it may be noted that industries based on oil and the internal combustion engine were hardly the monopoly of the United States prior to the mid-twentieth century: countries such as Britain, France, Germany, Italy, Japan, Czechoslovakia, Russia, and several others had well-established automobile manufacturing complexes and refineries prior to the mid-twentieth century; the Second World War is witness to this technological fact—it was not exactly fought on horseback. The more accurate perception would be that the twentieth century, as a whole, has been the century of oil just as the nineteenth century was that of coal. To remind readers yet again, Marx did not live to see the era of oil or the era of electricity, both of which freed the factory system of industrial capitalism from locational constraints and brought distant markets and colonies under the global control of industrial, commercial, and financial capital. Coal gave a tantalizing glimpse of those potentialities, oil and electricity carried them much further toward realization. This is central to understanding the climate change and the environmental crises in the twenty-first century, to which we shall refer later in this monograph.

THE PERIOD 1991–2015: THE RISE OF GLOBAL NEO-LIBERALISM

The last decade of the twentieth century and the early part of this century has witnessed the neo-liberal counter-revolution across global capitalism. Socialism was globally on the back foot after the collapse of the USSR and the near collapse of East European economies which had adhered to the central planning model. The combined political and economic onslaught on the countries, particularly of the Third World, has been very well described by David Harvey (Harvey 2007). In brief, an unrestrained capitalism went on a global rampage, spearheaded by the United States and EU. Third World countries were forced to open their markets to global investment capital and commodity trade, to change their national laws toward this end, to reform their economic structure and institutions, to dismantle state-owned enterprises and wind up state welfare programs, to accept international (that is, capitalist) "conditionalities" in redesigning their economies, and so on. Not to put too polite a gloss on it, but this has been a period of economic imperialism without having to physically occupy or administer entire countries, merely draining them of their economic surplus through the exploitation of cheap labor working in "indecent" conditions; the off-shoring of environmentally polluting and damaging production processes, inducing an economic "race to the bottom" through competition among poor countries and sometimes even between different states within a country; and so on. This period has also seen the emergence of the World Trade Organization (WTO), the third leg of the triad of the institutions proposed by Keynes who had referred to it as the International Trade Organization (ITO), the other two legs being the World Bank and the International Monetary Fund (IMF) referred to earlier. This third leg of the Keynesian triad has permitted the completion of the circulation of capital in its interrelated forms of money-commodity-money or commodity-money-commodity which Marx had outlined in *Das Kapital*. This has also been the period which has seen phenomenal economic growth in China, catapulting it to second place among nations in the global economy in terms of size. Over a period of two decades or more, the average growth rate of China at over 10 percent per annum (that is, doubling every seven years) has perhaps never been witnessed before in the global economy. The Chinese economy has not only acted as a "locomotive of growth" for the global economy, it has also pulled nearly seven hundred million Chinese out of historic levels of poverty and converted China into a middle-income country in a matter of one generation. This performance has not been lost on other developing countries across the world. China's rise to prosperity has largely been powered by coal and petroleum fuels, at a time when global knowledge of the climate crisis has exploded across the international stage. The rise of China is now being perceived as a threat to capitalist hegemony globally, led by the United States in particular. During this period, the Japanese economy has been in the doldrums and the Japanese have joined the United States in their concern regarding the global shift of power. Furthermore, after nearly a decade of near economic meltdown, Russia has re-emerged as a military and petroleum power on the world stage, further recasting the power equations in the global arena.

This necessarily compact narration of history, starting from the French Revolution and the Napoleonic wars and ending in the early decades of the twenty-first century, may enable us to acquire the needed perspective to begin a reasoned analysis of the crises confronting global humanity in the twenty-first century.

Chapter Four

Existence-Threatening Crises Confronting Global Humanity in the Twenty-First Century

To recapitulate:

In the preceding sections, we have combined a critical view of history, as pertaining mostly to Europe and by extension to both America and Russia, along with a critical overview of the political economy of capitalism as a system. In attempting to develop the view along two inter-related themes, namely, the political revolution in France and its impact on European history and political economy in the nineteenth century and the Industrial Revolution across Europe both in the nineteenth and the twentieth centuries and its impact across the world, we have tried to keep in sight the major crises facing the world in the twenty-first century. These crises are the nuclear arms race or the crisis posited by a generalized international arms race, the generalized crisis of the global environment, and the climate crisis. There is a debt owed to Karl Marx, the public intellectual of the nineteenth century without whose untiring critical labors we might have missed the clarity of analysis of both the intertwined revolutions of the nineteenth century—the political and the industrial—and the class interests driving or thwarting them. Indeed, Marx's writings and arguments are difficult to understand without an adequate background understanding of the major event—structures of nineteenth-century Europe—hence we have devoted some effort to outlining those major event structures. However, having embarked on the history of Europe in the nineteenth century, we have had to continue with it through the period of the twentieth century and the first decades of the twenty-first. With this as a background, we now turn to an analysis of two of the major crises: the international arms and race and the environmental crisis. Our understanding of the third major crisis—the climate crisis—will be more fully developed later in the course of this monograph.

THE CRISIS OF NUCLEAR ARMS: FAST DEATH

The nuclear arms race was inaugurated by the use of nuclear weapons dropped on Hiroshima and Nagasaki toward the end of World War II. This is the only instance in world history hitherto in which nuclear arms have actually been used in war. Ever since, however, the world has lived under the shadow of nuclear weapons of mass destruction, as countries have scrambled to acquire these weapons: starting with the United States, followed by the UK, France, Russia, and China, and later others: India, Pakistan, Israel and now North Korea. Yet this begs the question: why would any nation wish to totally destroy the people of an enemy nation—what is it that drives and fuels the enormous hatred of other peoples of the world, to the point of the threat of complete annihilation?

One advantage of looking at European history starting from the French Revolution and the Napoleonic wars is that it may shed some light on this wretched question. At the risk of offending the purists among the historians, we may consider the Napoleonic War as World War Zero. At its peak, the Napoleonic empire stretched from France along Germany and Poland up to Moscow; on the southern side, it encompassed Spain and large parts of Italy; on the west, Netherlands and Belgium; and toward the north, parts of Denmark. Naval battles were fought along the coasts of Western Europe and the Mediterranean as well as North Africa and land battles in North Africa. The effort to curb the spread of the Napoleonic empire resulted in cutting the links of Spain and Portugal with their colonies in South America. Moreover, British and French troops fought on Canadian soil as well as on the Indian sub-continent. It is in consideration of the geographical spread and its global impact which makes us rethink whether the Napoleonic War should not be reconsidered as World War Zero.

As indicated earlier, the conservative powers of Europe allied themselves against Napoleon and these conservative powers included royalty, aristocracy, and the Church, so much so that the various branches of the Church in Europe—Catholic, Protestant, and Orthodox—sank their difference in the face of the threat to their class. The royalty and aristocracy of Europe were dynastic hereditary kingdoms and empires, related to each other by centuries of inter-marriage, their power sanctified by the Church and the doctrine of the "divine right of kings." These divinities in human form and shape enjoyed absolute power—the power of life and death—over their subject peoples and they were unwilling to part with that power, which was threatened by the French Revolution and the republican promise of self-rule by representatives of the people. It has been said that "power corrupts; absolute power corrupts absolutely"; in line with this saying then, the absolute power exercised by the ancien régimes of Europe represented their absolute corruption. It is only such absolutely corrupt regimes that could generate the venom that fueled the repression across Europe in the nineteenth century and was carried over to the wars of the twentieth century. Absolutist regimes answerable only to God can perform any atrocities on their subjects, as history repeatedly shows. And if they can unleash atrocities on their own subjects, they can unleash even greater harm on others and attribute it to fulfilling God's will.

This understanding of wars fought in the name of religion is important because religious wars never seem to end: the Crusades were fought for over a century, the Catholics and the Protestants fought in Ireland for over three hundred years even though the Thirty Years' War fought between Catholics and Protestants across northern Europe (1618–1648) ended in the Treaty of Westphalia, a landmark event in European history; the Jews have been persecuted for centuries across all of Europe and Russia, the Shias and the Sunnis have been at each other for over a thousand years; instances of other wars and massacres in the name of religion across the world can be multiplied. When contending religious forces, each claiming absolute truth for their beliefs, confront each other, the issues can rarely be settled through reason which is a child of modernity. Moreover, confrontation in the guise of religion tends to consolidate the hold of the conservatives over the rest of society and these in turn stoke memories of real or imagined atrocities perpetrated on their followers by the other side, ensuring the transmission of hatred from generation to generation. In comparison with the religions of the world, the triad of rationality-science-reason is but a child which is a few centuries old; yet it has made its mark, particularly in the twentieth century. Most importantly, enquiries based on reason and evidence often tend to discover material interests underlying the sanctimonious humbug projected in the name and cause of religion. Converting therefore a nuclear-armed confrontation into a quasi-religious herd mentality is a step backward into a pre-modern world fraught with unreasoning anti-intellectual attitudes and pre-suppositions which pose a danger to humanity at large. This is what the mind-set of the Cold War years represented.

The Cold War may have officially ended; that does not mean that the mind-set has been erased, as though by pressing an electronic button an entire slice of historical memory can be wiped out. The modern world owes much to one man—Mikhail Gorbachev—than the twentieth century has been willing to acknowledge; perhaps the twenty-first century will be more gracious.

> That is why the world owes so enormous a debt to Mikhail Gorbachev, who not only took this initiative [of ending the Cold War] but succeeded, single-handed, in convincing the US government and others in the West that he meant what he said. However, let us not underestimate the contribution of President Reagan whose simple-minded idealism broke through the unusually dense screen of ideologists, fanatics, careerists, desperados and professional warriors around him to let himself be convinced. For practical purposes the Cold War ended at the two summits of Reykjavik (1986) and Washington (1987). (Hobsbawm 1994, pp. 249–50)

However, Hobsbawm goes on to further say that

> it could not be universally recognized as being at an end until the USSR had visibly ceased to be a super power. . . . It was the collapse of the Soviet empire in 1989, the disintegration and dissolution of the USSR itself in 1989-91 which made it impossible to pretend, let alone to believe, that nothing had changed. (p. 251)

It is worthwhile to ask, a quarter century later, whether the dragon's teeth of the military-industrial complex, sown during the Cold War, do not continue to provide a bitter harvest. At its peak in the eighties, 7 percent of the US GDP was devoted to war outlays. At the end of the Cold War, a so-called "peace dividend" had been promised across the United States to its citizens. We do not believe that promise was fulfilled; it was probably never intended to be, such being the politics of those who control the military-industrial complex.

Even at the start of the twenty-first century, the US military budget at around USD 700 billion is probably larger than the combined military expenditures of the next ten countries in the world. Despite various nuclear arms limitations agreements, the world of the military super powers remains awash with enough nuclear weapons to destroy life on earth several times over, if not through a "nuclear winter" then almost certainly through the radioactive fallout which will spread globally through air and water currents (as in Fukushima magnified) since some of the radio-nuclides have half-lives of tens of thousands of years. Another legacy that the twenty-first century will inherit is the nuclear fallout from weapons testing—US and French—in the Pacific South Sea islands. Our recent mapping of ocean currents, as part of the climate science, indicates the presence of a great conveyor belt of ocean currents virtually circling the globe, which means that radioactive fallout from weapons testing would now have reached the United States and Europe. It is a strange reflection on democratic politics that we elect leaders to terms of office that may extend to four or five years but they can take decisions whose consequences will extend irreversibly for all life forms across earth, for tens of thousands of years. The millennial generation would do well to ponder on this question, as would the various environmental movements across the world, as it intersects two of the major crises confronting us: nuclear war and the global environment.

There is another intersection: between the war machine and the climate crisis. The US Pentagon is the largest single consumer of petroleum in the United States and therefore probably in the world. So long as they can commandeer monies from the public budget of the United States, this climate destructive behavior is unlikely to change. The United States has between 900 and 1,000 military bases on foreign soils—we are thankful to Noam Chomsky for having brought this to light; no other country is even remotely close to this figure when it comes to projecting military force internationally.

There are other intersections between the three major crises: nuclear weapons, environment, and climate. One of the most important of these intersections is between nuclear weapons and civilian nuclear power, both of which share a common fuel: enriched uranium. Nuclear weapons use uranium enriched up to 85 to 90 percent for fission devices whereas nuclear reactors use 5 percent enrichment. The enrichment technologies are common; very often, so is the pool of highly trained personnel, hence the nuclear weapons lobbies will support the civilian nuclear power lobbies as they would like to keep intact the pool of trained personnel in order to be able to draw upon them when necessary. Civilian nuclear power is at least three times as expensive (per kilowatt-hour) as RE based electricity today (Raju and Ramana 2013, p. 58), yet orders are being placed in both India and China for western-manufactured nuclear power reactors. It may be noted that, as usual, these prices do not factor in the costs of decommissioning and the costs of long-term storage of nuclear wastes, both of which if factored in would make the price of nuclear power even higher. These nuclear wastes will leave behind yet another huge and dangerous environmental liability for future generations. This highly expensive nuclear power (dubbed "clean" and "non CO_2 emitting" hence climate friendly, both scientifically incorrect and absurd assertions) is taking away large chunks of the future electricity market from lower-cost climate-friendly renewables, yet a powerful combination of military and civilian interests continue to provide a lifeline to a dying nuclear power industry globally. Further, since such expensive electricity cannot be sold directly on the market, it will draw upon huge state subsidies therefore diverting financial resources from far better climate-friendly objectives.

In democratic societies, the easiest route for breaking this unholy nexus is to clamp down on defense budgets of the "national security" state. Internationally, nuclear weapons need to be banned, as chemical and biological weapons have been via the Vienna Convention with no exceptions. Due to the link between the military and civilian nuclear industry, it is also necessary to bring to closure the nuclear power industry worldwide with no exceptions. As an example of the negative interaction between defense and "national security" expenditures on the one hand and social welfare expenditures on the other, one may look at India's MGNREGA (Mahatma Gandhi National Rural Employment Guarantee Act) enacted by a previous government and centrally funded by it. This act guarantees every rural household in India a minimum employment of one hundred days per year on demand, as an economic right. The World Bank estimated a requirement of 1.7 percent of India's GDP for adequately funding the program, yet at its best the central government has only managed 0.58 percent of GDP. This is probably the largest social welfare program in the world in terms of number of persons covered, yet the present right-wing government in India has brought down the allocation to a mere 0.30 percent of GDP in 2016–2017, virtually destroying the program and therewith the rights of the poor in India (Patnaik 2019). This program was arguably responsible for the decade of high economic growth witnessed in the Indian economy yet it is being starved of funds even as huge funding is being made available for defense purchases, unconscionably high-cost nuclear reactors for electricity, and so on. We shall have more to say about such expenditures in a later part of this monograph.

Even though nuclear weapons and the missiles to launch them form a significant and dangerous component of the war arsenal of super powers and aspirants, they are by no means the complete set of weaponry. Earlier, we had indicated that by the end of the Second World War, the theater of war had already become four dimensional, that is, land, sea, air, and computer-based cryptographic communications. The huge advances in computer technologies, both hardware and software, have revolutionized warfare in the other three domains as well, with corresponding costs of regular upgradation. Expert opinion holds that the theater of war has now become five dimensional with the addition of space as the fifth dimension. Every major nation now has its own satellites and global positioning systems; we are already into the era of *Star Wars* with laser weapons to destroy satellites, anti-ballistic missiles to target enemy missiles, satellite-controlled drones, cyber warfare to interfere with communications, and control, and so on. These were considered techno-fantasies during the Reagan era, now become battlefield realities. The "ideologists, fanatics, careerists, desperados and professional warriors" identified by Hobsbawm have been busy indeed in creating ever more efficient ways of killing humans, singly or in large numbers. Instead of making the world more secure, the opposite objective has been achieved. Recently, the Union of Concerned Scientists which has been maintaining the Doomsday Clock has set it at two and a half minutes to doomsday, the shortest in about half a century including most of the Cold War period. This is the scenario we have chosen to name as Fast Death; it is maddening to any sane and civilized mind that huge resources are being channelized to such horrendously ignoble and perverse ends even as perhaps 30 to 40 percent of global humanity struggles to survive on less than US 2 dollars per capita per day. Even more to the point, the military expenditures acknowledged by governments around the world hardly reflect the full story since governments are quite adept at smuggling military expenditures under innocuous-sounding civilian budget heads of expenditure. In short, in democratic societies, most of us are in the dark about actual military expenditure in our respective countries, to bolster military-industrial Keynesianism. As indicated earlier, the only way to de-escalate the Fast Death scenario is to enforce a long-term curb on military expenditures of governments; no doubt, powerful lobbies will resort to fear-mongering to ensure this does not happen. Nor can the press be relied upon to take up the public cause, being too dependent on advertisement revenue and state goodwill. As an anonymous wag put it: the freedom of the press is the freedom of the editor to publish the views of the owner provided the advertisers do not object.

It is, therefore, somewhat of a relief that the worldwide web of electronic communications has made us less dependent on tailored opinions.

It is sometimes asked whether there is anything to be learned from the study of history. Since we have started with the Napoleonic wars in Europe, it may be educative to look at the political activities of the nasty Prince Metternich, the Chancellor of Austria during and after the Napoleonic period. Metternich

> proposed postal conventions that would enable foreign correspondence to pass through Austria. In Vienna, he set up a special office for opening, decoding and re-sealing all correspondence that came through the post. Everything of interest was copied and passed on to the Chancellery. This gave Metternich an astonishing amount of information about all foreign governments, and it was supplemented by constant reports from his army of spies and secret agents, as well as from the police. To know in intimate detail what was going on everywhere became something of an obsession with him. (Thomson 1974, p. 113)

It was no wonder that Marx never missed an opportunity to take a swipe at Metternich.

In our current context, it is known that all electronic correspondence worldwide is being stored by governments which have the means to do so—in other words, our taxes empower our governments to spy upon us. How different is that from the system of Metternich?

Defense, it has been said, is moral. By implication, it is offense that is immoral. Yet the world of the super powers, as we find it today, seems to believe that offense is the best form of defense; this nostrum blurs the dividing line between offense and defense, making it suitable to deploy aggressive doctrines that conveniently obfuscate the moral dimensions of war.

Some of the greatest figures of the twentieth century have been opposed to the use of nuclear weapons: Einstein, Bertrand, Russell, and Mahatma Gandhi, to name but a few. The latter part of the twentieth century did see an anti-nuclear popular movement in the western countries; the twenty-first century must ensure that these memories are not buried.

These are some of the systemic interconnections between nuclear weapons, the environmental crisis, the climate crisis, and the overall structure of the international system of political economy. More such interactions will be indicated in the further examination of the other crises.

THE CRISIS OF THE GLOBAL ENVIRONMENT: SLOW DEATH

There has been a significant body of Marxist literature on the subject of the internal evidence in Marx's writings linking the ongoing class struggle in society with the despoliation of nature, that is, what we call "the environment" in the second half of the twentieth century. This debate has emerged in the second half of the twentieth century parallel to the rise of environmental movements in the United States and Western Europe and gradually spreading to other parts of the world. Some of the environmental movements particularly in the United States (for example, the Sierra Club) predate the above timeline and have a very large membership, over a million members, but confined to the United States. Others, such as Greenpeace and Worldwide Fund for Nature, have acquired international membership. Most of them have started from being single-issue movements (for example, Sierra Club centered on forest/landscape conservation, WWF-N on species conservation, Greenpeace on industrial pollution and anti-nuclear campaigns, etc.) and operated within their own grooves though the very nature of their activism has confronted them with the other impacts involved in the concept "despoliation of nature" which at once demands a broader, more generalized understanding of nature along with an understanding of the specific impacts of human activities on nature. To be fair to environmentalism as a movement in society, however, it was a protest against not just the destruction of nature but equally against the appropriation and destruction of the "life-world" of all those brought forcibly under the dominance of industrial capital.

> Marx . . . [along with] . . . Keynes and Leontieff, the great theorists of the modern economy have all seen "disposable time" for activity that is "an end in itself" . . . and . . . as "the true measure of wealth." (Gorz 2018)

In other words, the citizens of the materially wealthy countries yet did not enjoy "true wealth" because they lacked "disposable time," the one entity that capitalism was unwilling to yield its control over. In his study of technology and its development during the nineteenth century in Britain, Professor David Landes had arrived at a strikingly similar conclusion:

> For the worker, the transformation was even more fundamental, for not only his occupational role but his very way of life was at stake. For many—though by no means for all—the introduction of machinery implied for the first time a complete separation from the means of production; the worker became a "hand." On almost all, however, the machine imposed a new discipline. No longer could the spinner turn her wheel and the weaver throw his shuttle at home, free of supervision, both in their own good time. Now the work had to be done in a factory, at a pace set by tireless, inanimate equipment, as part of a large team that had to begin, pause and stop in unison—all under the close eyes of overseers, enforcing assiduity by moral, pecuniary, occasionally even physical means of compulsion. The factory was a new kind of prison; the clock a new kind of jailor. (Landes 1969, pp. 42–43)

Where the good professor errs is in characterizing the clock as a jailor; the jailor is always a person, embodied within the factory system by a functionary nominated by capital to enforce its authority in the place of production—be it as overseer, supervisor, manager, CEO, or in general "boss" whose responsibility is not to "work" but to get the work done. Environmentalism would have benefitted greatly from locating precisely the locus of the destruction of the "life world" of the majority of the people since the rule of the clock has now become ubiquitous—from factory, to office, to shop, to commute, to bank, to government, and so on—this loss of "disposable time" as true wealth has assumed an overarching form governing all of social life under the dictatorship of capital. This insight into the nature of the capitalist system—that it is, in effect, a series of interconnected prisons—is capable of transforming our view of the capitalist system and therewith the ways of transcending it in its entirety since it robs its citizens of their precious freedoms.

However, along with the environmental movements, the above timeline also includes the rise of Green Parties in the various countries of Western Europe and the United States—parties which have entered mainstream electoral politics with a primary focus on green issues and have had varying degrees of success in greening the agendas of the two main political streams, the conservatives, and the social democrat/labor parties. However, despite occasional local successes, neither the parties nor the movements have made a significant dent on the destruction of nature that has proceeded hand in hand with the Industrial Revolution at a global level. Nevertheless, a significant body of non-Marxist literature on the environment has been generated by these movements and parties.

Recent academic literature has also used the term "eco-socialism" to describe the new green consciousness that has emerged from the environmental movements in the developed capitalist countries. While critical of capitalism and its evident failure to come to terms with the enormity of the global environmental crisis, most such writings fail to integrate the event structures of socialist societies or of the combined developmental and livelihood/environmental struggles in the countries of the post-colonial periphery, which together form a major part of the global population. To that extent, these debates have remained centered on the concerns of the developed capitalist world.

However, a sea change is now visible in historiography, for example, in the writings of Beckert:

> These frontiers of capitalism are often to be found in the world's countryside, and the journey through the empire of cotton reveals that the global countryside should be at the centre of our thinking about the origins of the modern world. (Beckert 2014, p. 441)

This is the reason why we have chosen to title this monograph *The Greening of Socialism* rather than to remain restricted to the limited debates emanating within the terminology of "eco-socialism."

As indicated earlier, the late eighteenth and nineteenth centuries in Western Europe witnessed two revolutions—the political revolution originating in France and the Industrial Revolution largely originating in Britain. In particular, Britain had already undergone a political revolution in the seventeenth century (the period from 1642 to 1688 was the period in Britain which saw the overthrow of the monarchy to be followed by the restitution of monarchy, however as a constitutional monarchy, so that Parliament was supreme in the making of laws, in the oversight of government and its finances, in the control of the armed forces, and so on [Hill 1970]). This paved the way for the later Industrial Revolution by converting the peasantry into a proletariat through dispossession. This early lead provided by the revolutionary upsurge in Britain was responsible for the lead that Britain developed in industrialization and the use of coal and the associated technologies to power industrial production. While most of the debates within Marxism concentrated on the political revolution and its aftermath, particularly in the Age of Catastrophe from 1914 to 1945, the Marxist debates on the environment arise from the aftermath of the second revolution—the Industrial Revolution—and the global despoliation of nature as a consequence of the spread of the industrial mode of production, that is, these later debates are embedded into a critique of industrialization under capitalism and presuppose an understanding of the earlier critique of political economy initiated by Marx.

In retrospect, that is, the wisdom of hindsight, it seems that the entire debate on the relationship between man and nature could not take off during a period when the entire imagination of the world was caught up in the crumbling of global empires, destructive wars spanning the entire globe, the drama of emerging socialist revolutions in Russia, China, and elsewhere, and equally dramatic emergence of newly liberated nation states in Asia, Africa, and South America as well as Eastern Europe. That debate had to await the relative peace prevailing during the period of the Golden Age of Capitalism and the Cold War. The writings of James O'Connor, Elmar Altvater, Joel Kovel, John Bellamy Foster, and Andre Gorz to name but a few of the leading voices in the ecological Marxist debates have thrown much light on the man-nature relationship. O'Connor founded the journal *Capitalism Nature Socialism* (*CNS*) in 1968 and after him, Joel Kovel assumed the editorship. The journal has now completed a half century of its existence, with contributors from all over the world, proving that this debate has struck roots across the world. Bellamy Foster, another prominent participant in the debate, has now assumed the editorship of *Monthly Review*, ensuring that between these two journals the debate will continue in future.

Within the ecological Marxists, the debate taking place is between two lines of argument. Briefly, O'Conner and Kovel have concentrated on building up the thesis of the second contradiction, that is, the man-nature contradiction, as being as important if not more important than the first contradiction as identified by Marx, that is, the man-man contradiction or the class struggle in society. According to the proponents of this thesis, the second contradiction remained largely implicit as a thread in Marx's writings, which concentrated much more on the emergence of the proletariat as a class which was the sworn historical enemy of the bourgeoisie, both classes being international; hence the need for the working classes across nations to ally with each other in international opposition to the bourgeoisie. On the other hand, Bellamy Foster and Andre Gorz (to an extent) trace the man-nature contradiction as explicated in Marx's writings to the "metabolic rift" between town and country, created by capitalism—in effect, the bourgeoisie as a largely urban class colonizes not merely its immediate countryside by extracting from it food, resources, and laboring populations but also using that countryside (and in a more generalized sense, the commons) as a dump ing ground for the wastes produced by industrial capitalism. In this thesis, the conception of "town" is generalizable to the metropolitan center, that is, the highly urbanized areas under capitalism, historically Western Europe and the United States, whereas the "country" is the peripheral countries of the world which under imperialism supplied the capitalist world with food, fibers, energy, materials, and cheap labor and continue to do so, tied into the capitalist system through a network of financial debt. The circulation of both capital and material commodities in fact created "ecological imperialism" not only during the expansionary and establishment phases of capitalism as a global system but in the very process of the emergence of capitalism through primitive accumulation. Marx is reported to have stated,

> The treasures captured outside Europe by undisguised looting, enslavement, and murder flowed back to the mother country and were turned into capital there. (Foster 2009, pp. 236–37)

In other words, the twin revolutions in Europe—political and industrial—were accompanied by a worldwide process of ecological imperialism wherein the treasures referred to by Marx included: land, forests, wildlife, plantation crops, minerals, energy resources, chemical inputs for agriculture, the plundered products of oceans and fresh water bodies, and, of course, coerced labor. It has required a half century period after decolonization for countries around the world to even begin to piece together the magnitude of the long-term ecological impact of colonialism on their countries in terms of the devastation of nature.

Bellamy Foster is correct in maintaining that

> we need to beware of falling into what EP Thompson called "the enormous condescension of posterity." More specifically, we need to recognize that Marx and Engels . . . had the advantage of living in a time when the transition from feudalism to capitalism was still taking place or had occurred in recent memory. Hence the questions that they raised about capitalist society and even about the relationship between society and nature were often more fundamental than what characterizes social and ecological thought, even on the left, today. (Foster 2008, p. 144)

Even more to the point, Bellamy Foster has unearthed the deep link between Marx's writings in *Capital* Volume I and Liebig's critical analysis of the "destructive side of modern agriculture" through a study of Liebig's scientific writings on agricultural chemistry, which also demonstrates Marx's thoroughgoing commitment to the worldview of the natural sciences, both physical and biological.

The two schools within ecological Marxism are broadly in agreement that (a) there is enough internal evidence in Marx's writings as regards recognition of the complex and mutually interactive relationship between society and nature under capitalism and that this recognition is highly critical of capitalism and (b) Marx's environmental critique of capitalism therefore predates later "environmentalism" by at least a century and (c) that both schools have had to contend with the usual obfuscatory noises generated by bourgeois debates on nature and the environment but also from orthodoxy of interpretation both within Marxism as well as within the broad left tradition of the critique of capitalism as system.

The differences in the two schools arise from the differing views of the conceptual "models" within Marx's writings regarding the relationship between society and nature under capitalism. The second contradiction school broadly maintains that much of the relationship between society and nature has been "implicit" in Marx's writings and that bringing it out in the open in theoretical terms is the task that is to be done. The metabolic rift school holds that Marx had explicitly identified the metabolic rift between town and country in his writings and this constitutes the core of his ecological critique of capitalism. To an extent, the first school holds a "linear" view of the interconnections between society and nature, the second school holds a "circular" view (somewhat similar to the circular steady-state view of the economy espoused by Herman Daly and John Cobb but also similar to the theory of the circulation of blood by William Harvey which Marx used in his analysis) of the society-nature relationship.

These perceived differences in the "models" should not be treated as doctrinal differences because both schools are completely in agreement that capitalism as a system poses a threat to the very survival of human society through its accelerated destruction of nature. In the natural sciences, if there are two "models" being offered, then a series of experiments can be undertaken to verify the respective claims to truth or validity but this is not possible in the social sciences—in other words, it is only in the social practices that constitute the social praxis for changing the society-nature relationship that the answers are likely to be discovered. In the countries of the periphery, that is, the Third World, the opposition to the expansion of capitalism through struggles over the environment actually transform into struggles over livelihoods, control over natural resources, opposition to dispossession, opposition to privatization, and so on and the range of environmental resources at the heart of such struggles range from land to forests to water to marine resources and so on.

The core of the second contradiction conceptual model can be viewed as similar to the double-helix strand of the DNA model in which the two strands of the double helix are bound to each other through the four nucleotides—adenine (A) always bonding with thymine (T) and cytosine (C) always bonding with guanine (G). This may be a useful way of thinking about how one strand of the helix man-man or society (first contradiction) is linked to the other strand of the second helix man-nature (second contradiction). The specific links which bind the two co-evolutionary strands together are what we now call by the generic term "technology" but in earlier epochs would be covered by the term "social practices of production." In tentatively proposing such a conceptual model, we may note that it makes distinct three different categories of current human knowledge:

- that about present day society, the study of which is the proper domain of the social or human sciences: history, political economy, sociology, anthropology, psychology, and social philosophy and their sub-disciplines and overlap

- that about nature as consisting of living entities, the study of which is the proper domain of biology and both its sub-branches botany and zoology but also of human and animal medicine and their sub-disciplines which include the study of wildlife in their natural habitats; and other disciplines such as microbiology and ecology which specifically brings together abiotic and biotic factors linking living organisms to their environment; agriculture should properly belong to this category of knowledge since this is the category that deals with all living entities
- that about technology which deals with the processes created by humanity for meeting all its needs of materials and energy through the industrial mode of production; the branches of which range over chemical technology, mechanical engineering, electrical engineering and electronics (including software), civil engineering, metallurgical engineering and their sub-disciplines and overlaps. Almost all of technology today is related to prior conceptual breakthroughs in the sciences including physics, chemistry, and lately biology in the form of genetic engineering.

In using this double helix as a way of thinking about underlying relationships, we may note the model is biological and not mechanistic and permits mutual interaction and feedback between the parts, as in any biological or ecological system; and, as in the case of DNA, it is capable of co-evolutionary change (or even breakdown under certain circumstances). Most importantly, for our present purposes, it permits us to isolate and concentrate on technology.

Instead of proceeding in the abstract, we will take a real-life environmental struggle in India between a gigantic multinational corporation and a local community which spearheaded the environmental opposition, a typical stand-off between David and Goliath.

Sneak preview—there is a happy ending.

FROM ABSTRACT TO CONCRETE: *DAVID VERSUS GOLIATH—A CASE STUDY*

Goliath in this case was the Pohang Steel Corporation (POSCO), a giant South Korean steel conglomerate which intended to establish a twelve million tons per annum steel plant in the State of Orissa in India. Orissa has very rich mineral reserves; it also has extensive and excellent forests, a significant tribal population, a low GDP: all the preconditions that attract the avarice of international capital, which holds out the promise of creating employment, adding to the state coffers through mining royalties and taxes, and so on. Goliath intended to invest USD 12 billion into a combined mining and steel mill project with its own electrical power plant based on coal. This was touted as the single largest foreign direct investment project in India, a little over a decade ago. Goliath intended to mine the iron ore in a captive mine, transport the ore, and then convert the ore into steel in India and export the entire steel to South Korea (though some later accounts claim that the finished steel products or part thereof would be sold in India). Government at both the state and central levels were keen to showcase the success of the project in order to attract further capital investment, hence were willing to offer a range of subsidies at both state and central levels. In other words, the decks were stacked against the local community.

David, in this case, was the local community consisting of tribal populations who stood to lose cultivable land and forests; there were also local non-tribal small landholding cultivators who stood to be dispossessed through acquisition of their lands which were rich agricultural lands. Other sections of the local community were not even aware of the impending danger to their well-being and livelihoods due to the many environmentally damaging consequences of the proposed investment. As in all such cases, David was extraordinarily weak. The local community had been organized into opposing the project by a local branch of the communist party. The principal lines of opposition were opposition to land acquisition, opposition to undervaluation of good quality agricultural lands, and so on—not at all capable of being built up into a strong platform for mobilization, nevertheless, physical opposition to forcible land acquisition was undertaken by the local community, and the use of coercive tactics by agencies operating on behalf of Goliath resulted in violent local confrontations which were then reported first by the local press and then by sections of the national press, which barely reflected the overall picture within which the struggle was taking place. Nor was the local branch of the Communist Party equipped to make a full projection of the damaging consequences to nature and society arising from the project, or to hire and pay for highly priced "consultants" to prepare the case on David's behalf.

Indian environmental legislation required Goliath to prepare an Environmental Impact Assessment (EIA) Report, to be a public document. Following publication of this document on government websites, all those to be affected by the project were legally entitled to state their objections at a public hearing which formed part of the procedure; the public hearing was chaired by a public official and a record of the proceedings to be submitted higher up the government ladder, both state and central, for subsequent decision. This EIA Report was the only publicly available information on the project; it was to be prepared by Goliath at his own cost, normally undertaken on Goliath's behalf by a highly paid environmental consultant whose responsibility was to migrate the project through the government channels up to the receipt of the final environmental clearance, accorded by the central government in the case of large projects. Needless to add, no environmental consultant, paid a large fee by a corporation, is likely to arrive at an adverse conclusion regarding the environmental and social impacts of the project. The only other project document which provides detailed overview of the project is the Detailed Project Report (DPR) which is a techno-financial document prepared by yet another highly priced consultancy firm. The DPR normally contains the relevant quantities and technical details along with financial costs and financial projections to establish the techno-financial viability/profitability of the project over its entire lifetime—say twenty-five to thirty years. Such a document is needed to raise borrowings from banks and equity investors to finance the project but this document is not a public document. Under the prevailing ethos favoring Goliath, such a document is considered a "commercial secret." In effect, very little information regarding the intentions and effects of Goliath's investment was available in the public domain, putting David at a tremendous informational disadvantage. Hence the EIA Report acquired overwhelming public importance.

India's code of environmental laws, drawn up over four decades or so, consists of six legal enactments: three laws governing the three physical states of matter (solid, liquid, and gaseous) and three laws governing biological states: forests, wildlife, and biodiversity protection and conservation. The three laws covering the physical states are the Prevention and Control of Water Pollution Act, the Prevention of Air Pollution Act, and the Solid Waste Management Rules under an omnibus Environment Protection Act; the three laws covering biological states are the Forest Protection Act, the Wildlife Protection Act, and the Biodiversity Act. Most countries would have a similar environmental code. The consultant had to ensure that none of the six environmental legislations were being violated. David, on the other hand, had to conclusively prove that any one or all of the six environmental acts stood to be violated by the project proposed by Goliath; hence environmental clearance was to be denied to the project. This was David's only chance to ensure that the pellet from his slingshot hit Goliath squarely between the eyes.

Goliath's environmental consultant produced an EIA Report of several hundred pages, based on the strategy of "suppressio veri, suggestio falsi," that is, suppress or obfuscate the truth and create a smoke screen of falsehoods to divert the attention of the authorities. Public authorities all over the world, both political and administrative, in countries both capitalist and "socialist," are similar to the government of Saudi Arabia: if somebody is able to bring in money, organize production for their own profit, and ensure a steady stream of revenue to the government over a period of thirty years, the public authorities are only too happy—that essentially is the nature of governance all over the world.

Goliath's consultants went about their task by considering only the first phase of the project (four million tons per annum of steel output, to be later expanded in a second phase at an unspecified future date to the full 12 million tons per annum), so that only a smaller quantity had to be placed on the public record for environmental impact assessment. From the environmental study, they excluded four aspects—the iron ore mining ("the Mining Project"), the transportation corridor for carrying iron ore from 200-kilometer distance to the site of the steel mill and the finished steel products to the port and beyond for export ("the Transportation Project"), the water pipelines over 87 kilometers long for carrying water from two major rivers, the Brahmani river for mining related consumption and the Mahanadi river for steel production ("the Water Project) and the townships at both locations—the mines and the steel mills ("the Township Project"), by trying to argue that these four "Projects" were separate projects for which separate EIA studies/reports would be filed at a later date. In effect, the only aspects to be considered were 4 million tons per annum of steel production and 1,300 megawatts of coal-based power with source of coal unspecified. The rest of the EIA was filled with irrelevant trivia concerning common or garden variety of local flora and fauna and how they would not be adversely affected. There were glaring omissions regarding threats to major forest, marine areas, and social wellbeing.

David made a mere seventeen-page written submission, in multiple copies as necessary for government purposes, which argued along the following lines for deriving the environmental and social impacts of Goliath's project:

a. that environmental impact assessment for all six projects had to be undertaken, together since they constituted one single project for mining 20 million tons of iron ore per year linked to the production of 12 million tons per annum of steel for which 1,300 megawatts of coal-based power was to be produced.
b. the sum total of solid waste produced on an annual basis (steel slag plus fly ash) was computed at approx 10 million tons per annum or 300 million tons over the life cycle of the project; this did not include the mining overburden to be disposed off while excavating 600 million tons of ore over the project lifetime. These quantities had to be accounted for under the Solid Waste Management Rules of the Environment Protection Act.
c. the sum total of water requirement on an annual basis for all consumption purposes—steel making, power generation, and township consumption—was computed at approx 140 billion litres per annum, which would completely pollute one major river, the Brahmani, from mine tailings for 200 kilometers, depriving all other riparian human communities from access to drinking water and water for irrigation and food security; the other major river, the Mahanadi (the name means "great river"), was already over-exploited and would be ren-

dered completely dry downstream. These major impacts had to be considered under the Prevention and Control of Water Pollution Act.

d. the sum total of all gaseous emissions—from blast furnace operations and power project operations—was computed at about 34 million tons of CO_2 per annum or 1 billion tons over the lifetime of the project; this constituted over 2 percent of India's then total annual CO_2 emissions of 1,500 million tons per annum. This had to be considered under the Air Pollution Act as at least four major coastal cities of Orissa were likely to suffer health consequences among the urban population if smokestack emissions were carried 250 kilometers downwind of project location. The climate impact of releasing 1 billion tons of CO_2 had also to be factored in at the national level.
e. the release of 34 million tons per annum of gaseous effluents from smoke stacks would destroy thousands of square kilometers of good quality forests due to acid deposition effects: this had to be considered under the Forest Conservation Act and the Wildlife Protection Act.
f. The combined effects of water pollution and air pollution and enhanced port activity had to be seriously considered since the large breeding grounds of the Olive Ridley Turtle were fairly close to the project site. This had to be considered under the Biodiversity Act.

David was thus able to argue that all six of the major environmental laws would be seriously violated by Goliath's project. In addition, there would be long-term lock-in of CO_2 emissions which would cause global damage. Finally, all this environmental and social damage was being subsidized through various government subsidies, so that even government revenues in future had to be offset against loss of foregone revenues which is what the subsidies represented. Moreover, South Korea would beneficially use the steel produced whereas India would bear the environmental and social damages, a highly embarrassing fact for governments at both central and state levels.

David could have made the environmental and social damages (the "externalities") of Goliath's project more explicit by modeling the gaseous smoke stack emissions and the liquid effluents from the various processes and linking them to specific damages be it in the domain of public health, damage to riparian and marine eco-systems and livelihoods, damage to forests, and so on; this would have required far more access to data and knowledge resources and time than, as is typically the case, David was in a position to command. Instead, by using Occam's Razor to strip away unnecessary layers of argument, David was nevertheless able to link causal chains to consequences and provide a reasoned basis to show the violation of the environmental code by Goliath's project.

Once these objections were recorded in the public legal procedures, the central government (Ministry of Environmental and Forests) ordered Goliath to produce an EIA document for all the diverse operations of the project. Various committees were appointed to study the environmental and social impacts; gradually, the tide turned. When the revised EIA was submitted, the government of India asked for sixty modifications to the EIA, twenty-eight for the steel plant and thirty-two for the port infrastructure. Goliath, however, was also busy. USD 12 billion of capital investment purchases a lot of political patronage including getting the South Korean president to pressurize the government of India to grant environmental clearance during his official visit to India. At the local level, apart from a media blitz, criminal cases were filed against activists from eight opposition villages—in short, all the usual tactics employed by capital, to splinter and suppress the opposition. But the majority of the eight villages held fast. Finally, after ten years of inability to move forward, Goliath threw in the glove.

Those interested in an opposite narrative may visit bizodisha.com, "Why POSCO Failed," dated May 17, 2017, by Vivek Pattanayak. A barebones outline of the complex event history is available at https://ejatlas.org/conflict/posco-odisha.

David prevailed. Goliath withdrew.

FROM CONCRETE TO ABSTRACT

To return once again from the concrete to the abstract or from the world of David and Goliath to our own earlier discussion on tentative models, the above case study does throw up some interesting reflections:

1. in the world of environmental struggles all over the world but particularly in the developing countries, for every one victory of David's there may be tens if not hundreds of victories for Goliath, which is a sobering thought in a world where capital is running amok.
2. in this case study, the second contradiction conceptual model or at least a double-helix form of it seems to be more insightful than the metabolic rift model, at least in furthering the struggle on the ground. The usefulness of the metabolic rift model will be discussed in a later section on climate change. As far as the case study is concerned, on the face of it the traditional working class does not seem to have played any role in it, except perhaps through the presence of the local branch of the Communist Party. Also if a theoretical model cannot be applied in practice, there is a break between theory and practice (praxis) which blocks the further evolution of both which is urgently needed. The task then is to expand our understanding of the concept "technology" used in the double-helix model.

In the real-life case study discussed above, "technology" is not used in the same sense as used by Goliath, who had made detailed calculations of process quantities and associated "costs" to ensure the profitability of his enterprise. Such detailed calculations (normally part of the Detailed Project Report or DPR) were needed not only for Goliath to check that his own equity investment (normally 30 percent of overall project cost) would yield a healthy return on investment, but to also assure Goliath's banker or financier (who would be expected to invest the balance 70 percent of the investment as an interest-bearing loan) that the loan was relatively risk-free and would be returned along with periodic interest in a reasonable period of time, say ten years from the commissioning of Goliath's project. Say Goliath were to take four years to establish the project: the next ten years would be devoted to repaying the banker his capital and interest. The next twenty years after that, Goliath would have been the full owner of the project after putting up only 30 percent of the investment, so that the return on equity in the later years would be very high, which is why Goliath was drawn to the project in the first place. Under these circumstances, neither Goliath nor his banker would be interested in taking on any "external costs," that is, the costs related to environmental and social damage.

Goliath and his banker expected "the state" to assume responsibility for the environmental and social "costs" imposed by the project, on the specious plea that the state earned revenues in the form of royalty on the mined ore and value added taxes on steel sold etc. notwithstanding that the state had already foregone future revenues through an array of subsidies accorded to Goliath's project: subsidies in the form of tax waivers, tax concessions, support measures, incentives, etc.—the term "subsidies" being scrupulously avoided. If, from the future revenues of the state from the project (say A per annum), the annualized subsidies (say B per annum) are subtracted, that is, A minus B, then the state would in effect hardly be getting any revenues but would be saddled with the huge and rising environmental and social costs in the future.

On the contrary, the term "technology" has been used by David precisely to calculate the quantities and types of solid, liquid, and gaseous effluents and by-products arising from the use of the specific technology—these are then physically, biologically, and socially linked to specific types of damages: to forests, to biomes, to land, to water, to agriculture, to human health, to social dispossession, and so on. These damages were not "monetized" as "costs," they were retained as physical, biological, or social damages, which it was possible to prove through inter-disciplinary sciences. By avoiding the "monetization" of "costs," David avoided the conceptual morass generated by the fictitious commodity "money" which, as discussed earlier, is part of the three fictitious commodities: land, labor, and money, created and propagated by capitalism. Moreover, the fictitious commodity "land" could be generalized to other public or common-property resources such as water, forests, biodiversity, marine breeding grounds, and so on. In other words, apart from land, all other common-property resources were either appropriable or allowed to be privately used for waste dumping at zero cost to Goliath's project. In this sense, Bellamy Foster is completely correct in characterizing capitalism as a system of unpaid costs. More importantly, as he has pointed out,

> this means overcoming, in a non-reductive way, the split between natural-physical science and social science that has been one of the main alienated intellectual products of bourgeois society. (Foster 2009, p. 212)

This observation is extremely important. Capitalism/ecological imperialism has succeeded over the last two centuries in its divide-and-rule strategy in the sphere of knowledge by separating the physical sciences (and technology) from the biological/life sciences and both these from the social sciences. While these "alienated intellectual products of bourgeois society" are busy competing rather than cooperating with each other, bourgeois or mainstream economics has managed to rule the roost; in the sphere of valuing the environments its main instrument seems to have been monetization of damage costs as if all the money in the world can bring back the dodo or can bring Ken Sarowiwa back to life. Money is a fictitious man-made variable whereas the physical and biological sciences deal with natural variables, often combining into irreversible effects based on proven natural laws. These cannot be equated, nor can natural laws be subordinated to the fiction of money which capitalism so assiduously propagates.

Avoiding the twin intellectual traps of alienated knowledge and monetization of damage "costs" remain two of the challenges to be immediately faced. As the real-life case study shows, linking the impacts of a production technology with its environmental and societal effects, within a framework of political economy deriving from ecological Marxism, can still prove effective in combating the global depredations of capital, both industrial and financial.

This view involves

> complex interactions among the three basic systems—the ecosystem, the production system and the economic system—that, together with social or political order, govern all human activity. The ecosystem—the great natural interwoven, ecological cycles that comprise the planet's skin, and the minerals that lie beneath it—provides all the resources that support human life and activity. The production system—the man-made network of agricultural and industrial processes—converts these resources into goods and services, the real wealth that sustains society: food, manufactured goods, transportation and communication.

> The economic system—the recipients of the real wealth created by the production system—transforms that wealth into earnings, profit, credit, savings, investment, taxes; and governs how that wealth is distributed and what is done with it.
>
> . . . logically, the economic system ought to conform to the requirements of the production system and the production system to the requirements of the ecosystem. The governing influence should flow from the ecosystem through the production system to the economic system. This is the rational ideal. In actual fact the relations among the three systems are the other way round. (Commoner 1976, p. 2)

What Commoner has referred to as the "production system" is what we have in mind when we use the term "technology" or in a more specific sense "technologies" that link the first and second contradictions. Some of the most important clusters of "technologies" that need the kind of methodological treatment as reflected in the above concrete case study are:

1. coal-based electrical power generation systems that link coal mines to power systems through transportation links and corridors (rail, road on shipping) and power transportation corridors; for the purpose of environmental assessment, a unit size of 1000 megawatts is suggested.
2. gas-based electrical power-generation systems that link gas fields to port infrastructure to gas pipelines to power stations; a unit size of 1000 megawatts is suggested.
3. petroleum refineries with all the downstream processing units, which may be in the hundreds, based on refinery products; together with the port infrastructure, pipelines, and product transportation corridors; suggested unit size 1 million tons per annum.
4. steel plants linked to iron ore mines and coking coal mines (as in the case study) with transportation infrastructure; suggested unit size 1 million tons per annum.
5. cement plants linking three types of mines—bauxite, coal, and limestone—to the cement production unit and the transportation corridors; suggested unit size 1 million tons per annum.
6. aluminum plants linking bauxite mines and power grids to the aluminium refinery along with transportation corridors; suggested unit size 1 million tons per annum.
7. large hydroelectric power stations, particularly their use of construction materials for dam building; suggested unit size 1000 megawatts.
8. large-scale paper production facility, together with its inputs—chlorine, bleach, electricity, biomass, and water—taken as a single unit; suggested unit size 250,000 tons per annum.

9. large-scale highway project—eight lane—with all construction materials but excluding bridges etc., which are landscape specific; suggested unit size 100 kilometers.
10. gas-based chemical fertilizer project, suggested unit size 1 million tons per annum.

The unit sizes suggested for analysis are scalable, that is, quantities can be derived by multiplying the unit size (of analysis) to the appropriate scale. Such types of capital intensive projects lie at the core of the expansion of industrial capitalism particularly across developing economies. We have excluded nuclear power projects since nuclear power needs to be phased out, as argued earlier. The ten core technologies suggested for environmental impact assessment fall into the following categories.

a. four of them supply the energy needs of the production system: three supply it in the form of grid electricity (coal-based power, gas-based power, and hydropower—in most countries around the world today, these three technologies would supply in excess of 80 percent of electricity needs); the fourth, the petroleum refinery, provides energy in the form of fuel for over 80 percent of worldwide transportation needs whether by road, shipping, or aircraft and partly for railways.
b. two of the technologies—steel and cement—are key infrastructure materials; both are energy intensive and support a wide array of transportation infrastructure including bridges, flyovers, runways of airports, as well as the growth of urbanization all over the world.
c. Aluminum is a highly versatile, energy-intensive material (a huge amount of electricity is consumed in the production of aluminum by electrolysis); it is widely used for electrical transmission, in the electronics industry, and such applications as aircraft.
d. the paper industry is one of the most chemically polluting industries using up huge quantities of biomass and water as well as energy and releasing multiple waste products into the surrounding environment.
e. large-scale highways provide the basis for the growth of the entire mass production of automobiles and use highly energy-intensive materials: quarried and crushed stone, concrete, steel, and asphalt.
f. gas-based chemical fertilizers are highly energy intensive and completely unnecessary in agriculture if carbon recycling is achieved via organic agriculture.

In the generic environmental impact analyses suggested above, it is not possible to directly factor in the environmental and social impacts, which are in every case location specific. However, it is possible, on the basis of material analysis, to indicate the kinds of environmental and social impacts that can be anticipated by communities affected by such projects. It is also possible, on the basis of material analysis, to test the impacts against specific environmental legislation, as in the case study above. Moreover, the analyses can also be extended to account for jobs created versus livelihoods lost or population displaced/dispossessed or whose health will be affected, on the basis of a cradle-to-grave life-cycle analysis. The life-cycle analysis of a project, by definition, has to extend over the life of the project, in most cases twenty-five or thirty years of operation. In case any of the pollutants (other than CO_2 which will be considered in the next major section on climate change) have potential impacts well beyond thirty years—such as persistent organic pollutants from agrochemicals in the environment, carcinogenic substances emitted or produced, etc., it may be argued that the twenty-five to thirty years damage assessment period is scientifically inadequate. Similarly, where pollutants may be carried hundreds or even thousands of kilometers away from a project site such as by pollution of flowing fresh water sources or marine pollution, a similar plea may be entered for extending the area of impact assessment. As indicated earlier, monetization of damages should be avoided as it is impermissible in accordance with the foundational definition and principles of ecology as a science. In any case, capitalism has enough mainstream economists at its command, some of them highly regarded professionally, to pervert every scientific argument by dealing exclusively with monetary valuations and discounting the future, downplaying damage "costs" especially if they affect "others," exaggerating "benefits" especially if they happen to be "economic" and beneficial to sustaining capitalist profits.

In one sense, the pure "environmental" argument against any single project, no matter how large, has a huge weakness, which is that every such project, when viewed from a global scale, can be shown to be "local"; in other words, spatial generalization and the interlinking of effects becomes difficult and remains fragmented in public consciousness provided the project is sufficiently far away, or affects remote "others" with whom "we" do not identify, or has dangerous consequences but the probability of "us" getting affected is sufficiently small to ignore, and so on.

Let us be frank enough to admit that there is a degree of callous self-centrism that is part of the dominant global ideology fostered by capitalism over centuries which has prevented public protests against the environmental depredations of the capitalist system. This is the reason that we have chosen to name the crisis as "slow death" as the expanding global capitalist systems chews away at both nature and society in remote areas through slow processes that are not considered globally threatening or "newsworthy." Examples of such processes would be: the overdrawal from the Ogallala reservoir which supports 20 percent of irrigated US agriculture; the pollution and eventual destruction of the Great Lakes and the St. Lawrence River; the drying up of the Colorado River before it reaches the ocean; the dead zone in the Gulf of Mexico of about 18,000 square kilometers; the death of the Aral Sea in the former Soviet Union (Barlow and Clarke 2004, pp. 9–46). Further enlightening examples can be found with respect to the Housatonic River polluted for decades with highly carcinogenic PCBs (poly-chlorinated biphenyls) by a world famous corporation and heavy metal pollution in the Upper Clark Fork River by another world famous mining corporation (Prud'homme 2011, pp. 32–42). Some of these are going to have such huge impacts over centuries in the future (examples from across the world can be cited) that they deserve to be ranked under a new category—Crimes against Humanity and Nature—yet such is the fragmentation of "alienated knowledge" even among well-meaning environmentalists that they refuse to see systemic generalization over global capitalism. This is akin to the blindness shown to the slow demise of the British peasantry in the seventeenth and eighteenth centuries ("the violence of gradualism"); it is also akin to the process which kept a majority of the population in the nineteenth and twentieth centuries in the metropolitan centers of capitalism ignorant about centuries of colonial repression in the peripheral countries; and so on. If such an alienated and fragmented discourse expects to save humanity and nature from the ravages of capitalism, it is a faint hope. There has to be a greater thrust at a systemic level, if "slow death" is to be averted; and effective proactive strategies developed before the damage occurs rather than incur clean-up and damage costs for centuries afterward. It is perfectly rational to state that prevention is better than cure, though this is not acceptable to capitalism as a system.

The larger environmental movements and organizations in the developed capitalist countries, which have the necessary expertise and human resources to create EIA formats for these ten clusters of technologies and publish them worldwide on the net, could perform this public service since these formats could then be accessed by communities around the world in their struggles. The communities often lack the requisite knowledge and resources and often the time to put together a comprehensive picture. More such clusters of technologies could be identified in the future, since this list is merely indicative. The wide reach of computer-based communication technologies makes it worth trying. Since these would be methodologically comprehensive formats which could provide qualitative and quantitative indication of environmental and social damage from specific clusters of technologies, the universities could help them through collaborative research across disciplines—technological sciences, biological sciences, and social sciences—and develop course curricula accordingly to promote discussion across disciplinary boundaries. Such methodologies would help communities to rapidly perform the task of large-scale integration across the continuum of space-time-biomes-people that is a necessary step ahead, if the struggles are to succeed. It may be useful to review Wassily Leontief's work on technology in this direction and to extend it. Moreover, such quantification may also be required to compute CO_2 emissions from various technology clusters around the world in future.

To revert to the framework of political economy, it is often held that shadowy finance capital drains surplus from entire economies, rendering them powerless and indebted to the desires and further manipulations of capital. This is true; most educated people around the world do not know how much capital has already been accumulated in the hands of "the 1 percent" or whatever other name they are referred to by. Professor of sociology Peter Phillips has spent decades tracking the growth of the transnational capitalist class worldwide and provides estimates of the amounts of investible capital that is now in the private hands of a very narrow class (Philips 2018, pp. 37–52). However, finance capital also links up globally with industrial capital for investment in the "productive economy" which, as is visible through the environmental debate, also destroys nature and existing social formations worldwide in order for capital to remake society in its own image. It is here that capital stops being shadowy and reveals its real face which is not productive but destructive. This is the face that future generations need to identify and checkmate before both nature and humankind are irreversibly damaged beyond a point of no return.

One of the important lessons to be derived from the theoretical debate between "pure environmentalism" and the critique of capitalist political economy which we have referred to as ecological Marxism is the debate regarding impermissibility of the monetization of environmental damages. To accept that money or monetary valuation techniques can account for and pay for environmental damages is a fallacy of the highest order. It not merely equates apples and oranges, it pretends that apples can be blithely equated with nuts, bolts, screws, and widgets, the favored terminology of many mainstream economic debates. Sound ecological science need not fall prey to this fallacy because once the barn door is opened to this fictional commodity named money, the entire merry-go-round of circular arguments begins: not merely about how much to pay but who pays and who gets paid; as if by some magical property of money, the payment of money by one human to another human (or one set of humans to another set of humans in a more generalized sense) can somehow alter the laws of ecology or natural laws in a larger sense of the term. The whole debate about money is a huge diversion and one, moreover, that has been carried forward into the debate on climate change and the future of humankind.

Another important lesson for "pure environmentalism" to derive from the debates within ecological Marxism relates to the creation of fictional "property rights" in common property resources—water, forests, biodiversity, minerals, airwaves, marine interface areas, the atmosphere and so on; just as land is a fictional commodity (a commodity, by definition, is that which can be produced for sale—land cannot be "produced," it is appropriated and privatized through the use of state power, hence it is a fictitious commodity), so too are the other common property resources also new forms of fictitious commodities in which a slew of new property rights are sought to be created in the name of "economic reforms." In essence, this is the transformation of common property into private property, leading to its eventual destruction through capitalist over-exploitation. This is a central systemic feature of the history of capitalism as a system; as the last few decades of worldwide neoliberalism have amply demonstrated, this agenda has been ruthlessly pursued by expanding capitalism worldwide. "Pure environmentalism" cannot counter this central tendency of capitalism, because, by and large, it has been in denial of the systemic tendencies of capitalism by falling prey to the money illusion generated by capitalism. Again, this understanding has to be transferred to the emergent arena of climate change where the same fallacious arguments have found root and continue to delay the recognition of the grave existential threat posed to humankind and the natural environment wherein money-based arguments are considered to trump all other arguments including those based on good science. If this argument is taken forward, it implies that mainstream economics is the greatest denier of climate change; therefore, the denial of climate change is far more deeply rooted in society than most observers are willing to concede. In short, "pure environmentalism" has proven its inability to connect the dots which point to the worldwide need for systemic change if a sixth extinction and all its drastic consequences are to be avoided. This then is the existential choice that confronts future generations: accept the capitalist system as it is and commit to future extinction or change the system, with all the uncertainties that it implies for the future.

Before closing the discussion on the environmental crisis, an insight may be obtained from the observation of a non-Marxist political economist, Lester Thurow:

> If you look at countries that are interested in environmentalism, or at the individuals who support environmentalism, one is struck by the extent to which environmentalism is an interest of the upper middle class. Poor countries and poor individuals aren't interested. . . . As our incomes rise, each of us shifts our focus of demand for more goods and services. Initially, we are only interested in physiological survival. Food constitutes our main demand. As we grow even wealthier, our demand shifts towards roomier housing and higher quality food. . . . Suppose now that a family has reached an economy level where they can afford good food, housing, vacations . . . all those goods and services that represent the American dream . . . what is there left that can mar their economic happiness? . . . Now they run into environmental pollution. . . . Environmentalism is a demand for more goods and services that does not differ from other consumption demands except that it can only be achieved collectively. . . . From this perspective, environmentalism is a natural product of a rising real standard of living . . . for many Americans, the next item on

> their acquisition agenda is a cleaner environment. (Thurow 1980/1981, pp. 104–5)

Earlier, it was declaimed that man does not live by bread alone; today, perhaps, we can flip this around to state that man does not die due to lack of bread alone, he can also die from poor air, contaminated water, poisoned food, and so on. States have been reluctant to act on this aspect because it would hurt capitalist interests, but that is where a middle-class environmentalist movement and a historically rooted working-class movement can seek common cause—the worker, after all, also consumes air, water, and food. The rich may be able to afford filtered air, bottled water, and chemical-free organic food but that is not true for the vast majority of the population even in prosperous countries—in effect, if they have to pay for these "luxuries," it is equivalent to an effective lowering of their incomes since necessities that were freely available yesterday have become pricey today—according to mainstream economics, "the market" decides. The situation in poorer countries is even more cruel, which is why we refer to this crisis as the crisis of slow death.

Apart from the discussions within ecological Marxism, the debate on the environment has largely remained a fragmentary debate. This is no longer true regarding the crisis of climate change which we deal with below, because climate change is now supported by a scientific-theoretical discourse that is capable of integration with the critique of political economy under capitalism.

THE CLIMATE CRISIS: CERTAIN DEATH

Superficially, there is an overlap between the effects of environmental degradation or destruction and the degradation or destruction resulting from climate change. It becomes necessary to delve into the "concept" of environment and the many different ways in which it is used.

In systems science, which was born around the thirties or forties of the twentieth century, the defining equation is deceptively simple:

S + E = U
or,
System + Environment = Universe

This is based on a slightly earlier, deceptively simple but profound mathematical theorem, to the effect that any closed curve divides a space into an inside and an outside.

In systems science, this closed curve becomes the system boundary which divides the inside (what is included within the system) from the outside (that is, not within the system, the environment). Soon after, Norbert Wiener gave birth to a whole new algebra by adding the concepts of input, output, and feedback to create the field of cybernetics which was rapidly developed, using the new algebra, in the new fields of electrical engineering and computer sciences, both hardware and software.

In systems science, defining the system boundary for any problem under consideration becomes one of the most important determinations because once defined, the system boundary cannot be arbitrarily changed until the analysis is completed. If, for any reason, the system boundary as defined proves to be incapable of delivering the necessary clarity to the problem, one can always go back and begin all over again with a redefinition of the system boundary followed by a fresh analysis. It may be noted that in any area of inquiry, the defining of the system boundary simultaneously delineates both the system and its environment, the meanings of which cannot be arbitrarily changed (that is, what is the system and what is its environment) without a redefinition of the system boundary. In the above formulation, it may be noted that "environment" is undifferentiated, consisting of all that is not included in "system." Wiener's contribution of "feedback" led to three, and only three, logical cases of connection between system output and system input via system feedback, namely when the feedback is zero (that is, no influence of output on input), when the feedback is negative (that is, when an increase in output leads to a decrease in input, which in turn results in a decrease in output, so that the system is self-regulating), and when the feedback is positive (that is, an increase in output results in an increase in input which further leads to an increase in output and so on, the system is not self-regulating, it is self-exciting or unstable).

Unfortunately, the notation "positive feedback" and "negative feedback" may have created some confusion in the public mind; in systems theory, positive feedback is generally destabilizing and negative feedback is self-regulating, quite unlike the usage of these terms in normal language where "positive" denotes that which is desirable and "negative" denotes that which is undesirable. To re-emphasize, in climate science, positive feedbacks are dangerous and negative feedbacks benign.

These considerations become important when dealing with climate science which defines the system boundary of the atmosphere and the other natural systems of the earth in a particular way at the starting point and therefore simultaneously defines other parts of "its environment." In other words, once the definitions of climate science are accepted, the "environmentalists'" earlier undifferentiated notions of environment have to be thought afresh, although many of them re-emerge in a more definitive set of interrelationships between different natural systems and the sciences related to them; for example, hydrology. They also lead to a different way of conceiving the relationship between natural systems and human society.

THE EMERGENT SCIENCE OF CLIMATE CHANGE

Almost 125 years ago, the entire framework for climate change arising from the heating of the atmosphere was proposed by a Swedish chemist, Svante Arrhenius, writing in the journal of the Royal Philosophical Society. Arrhenius went on to receive the Nobel Prize a few years later, in another field of chemistry, but it is likely that future generations will remember him as the founder of the science of climate change.

The World Meteorological Organization (WMO), part of the UN system, defined in 1975 the climate system as being composed of the atmosphere, the hydrosphere, the cryosphere, the land surface (sometimes referred to as the lithosphere or the geosphere), and the biosphere (McGuffie and Sellers 2005, p. 5). Most other definitions would be similar. The climate science states that the increase in the atmospheric concentration of CO_2 (and other greenhouse gases) leads to the "greenhouse effect" because incoming solar radiation gets trapped in the atmosphere due to these gases. The underlying physics is straightforward: incoming solar radiation arrives at the top of the earth's atmosphere as short wave radiation because it has been emitted by the sun, a very hot source. This short wave radiation is able to penetrate the CO_2 and other gases in the atmosphere because these gases are transparent to it. At night, the earth (that is, land and ocean surfaces) emit radiation back to outer space because outer space is at a temperature close to absolute zero (–273°C) but this radiation from the earth's surfaces is in the form of long wave radiation because it is being emitted by a source which is at a much lower temperature than the sun. The CO_2 (and other greenhouse gases or GHGs) is opaque to long wave radiation, acting as a blanket and preventing the radiation from leaving the atmosphere. In effect, then, there is an energy imbalance on a daily basis between the incoming radiation and the outgoing radiation and this daily imbalance accumulates as heat in the atmosphere. This accumulated heat in the atmosphere tends to move to other cooler systems in the overall earth system, again through the well-known laws of physics governing heat flows in solids and fluids, namely the processes of conduction and convection. If the different components of the earth system are taken in pairwise order, the following processes are triggered:

a. *Atmosphere-hydrosphere interaction:* here hydrosphere refers to all the water bodies on the earth's surface—oceans, seas, lakes, rivers, ponds, etc. The heat flow into the hydrosphere results in greater evaporation of water into the atmosphere; water vapor is a greenhouse gas, more potent in its radiation trapping property than CO_2. This is one of the most powerful "positive feedbacks" that contribute to instability in the climate system; it is referred to as "the intensification of the hydrological cycle." Knowledge of the hydrological cycle is generally part of the curriculum in civil engineering.

b. *Atmosphere-cryosphere interaction:* here cryosphere refers to all the frozen parts of the earth system including the polar regions—the Arctic and Antarctic, as well as glaciers worldwide including Greenland, the Himalayas, and Tibet (often referred to the third polar region of the world due to its extent) and other high altitude mountain areas across the continents of the world. The atmosphere transmits heat to the icepacks in these frozen areas, causing them to melt. The melt water on the ice packs absorbs solar radiation instead of reflecting it without change of frequency unlike the surrounding ice, so that the melt water heats up and, in accelerating fashion, melts larger areas of surrounding ice; this is known as the "ice-albedo effect" (albedo means reflectance) or the albedo flip wherein the incoming shortwave radiation, instead of getting reflected without change of frequency, gets absorbed by water. This is another "positive feedback" effect. The melt water from land-based icepacks eventually finds its way to the oceans, triggering an inexorable process of sea level rise.

c. *Atmosphere-lithosphere interaction:* the atmosphere transfers heat to the upper layer of the land, that is, the soil, drying it out and starting or accelerating processes of desertification. This drying adds water vapor, a potent GHG, to the atmosphere in another contribution to positive feedback. In higher latitude areas with a permafrost layer about 3 feet below the surface, the heating of the soil leads to release of methane trapped in the permafrost. Methane is a potent GHG, about twenty-one times more effective than CO_2 in trapping radiation, so that this is yet another positive feedback effect. Moreover, as students of soil chemistry and the agricultural sciences know, an increase in the temperature of soil increases the rate of oxidation of soil carbon, yet another positive feedback effect.

d. *Atmosphere-biosphere interaction:* here biosphere is conceived of as the interconnected web of life-forms on both land and the oceans. As the atmosphere heats up, most green matter on land (that is, trees, grasses, plants, bushes, creepers, etc.) will increase the rate of evapotranspiration to stay cool, thereby adding more water vapor to the atmosphere in a positive feedback process. Since the land is already drying due to direct evaporation effects, this then results in the drying of land-based vegetation and an increase in forest and grassland fires which contribute CO_2 to the atmosphere, another positive feedback effect. Most of the pairwise interaction effects are based on physics, with some understanding of biological processes, for example, photosynthesis. Some of the interaction effects also involve chemistry and biology.

e. As the heating of the upper layers of the oceans proceeds, the heat moves downward through the water. In certain locations, for example, the South Kara Sea on the Arctic coast of Russia, the sea bed has huge deposits of methane clathrates (methane trapped in a crystal lattice of ice due to a combination of low temperature and high pressure); it is feared that when the traveling heat front hits the clathrate deposits, it will result in huge release of methane to the atmosphere, a major positive feedback effect.
f. The increase in atmospheric CO_2 results in the formation of weak carbonic acid on the ocean surface layer; this weak acid interacts with calcium carbonate found in the shells and skeletons of various oceanic life forms; one of these interactions leads to the bleaching of coral formations worldwide. The role of corals is to act as sources of rich biodiversity in the oceans, rather like biodiversity rich equatorial forests on land.

The consideration of these pairwise interaction effects within the earth system indicates that it is full of positive feedback, that is, self-exciting effects, which leads climate scientists to characterize the climate system as an unstable system, sometimes referring to the atmosphere as the "nervous" part of the earth system. Naturalist Tim Flannery writes: "if we were to imagine Earth as an onion, our atmosphere would be no thicker than its outermost parchment skin" (Flannery 2005, p. 22).

In other words, while the radius of the earth is about 6,300 kilometers, the thickness of the major part of the atmosphere (which contains more than 90 percent of the atmospheric gases) is just about 10 kilometers above the surface of the earth, so the onion skin analogy is accurate. It is the gaseous composition of this extremely thin layer that has been altered by anthropogenic activity in the age of the Industrial Revolution.

How strong is the heating effect caused by the increase in atmospheric CO_2 concentration? Scientists have a reputation for nerdiness: one calculation shows that the net rate of heating of the atmosphere is the equivalent of four Hiroshima-sized nuclear weapons detonated every second, that is, 354,600 Hiroshima bombs every day (Holmes 2013) and (Cook 2013). One Hiroshima bomb apparently generates 6.3×10^{13} Joules of heat while the earth's rate of heating is estimated at 2.5×10^{14} J/sec. This is the rate of heating being generated in the parchment thin layer of the global atmosphere, with the rate increasing as the CO_2 concentration rises. It is not only the atmosphere that is nervous, the climate scientists are nervous as well, for very good reason.

The physical, chemical, and biological evidence of this heating is amply documented by various scientific institutions and observations/measurements around the world. These direct observations include: breaking of high global average temperature records almost every year over the past decade; increase of the melting of Arctic ice cap almost every year leading to ice area minima every year; large-scale breaks in Antarctic ice packs; increasing incidence of forest fires globally; increasing frequency and ferocity of cyclonic storm events; changes in natural habitats and rates of extinction of species at unprecedented rates; release of methane from permafrost in high northern latitudes; and so on. Evidence is available from the websites of reputed organizations such as the World Meteorological Organization (WMO), the National Climate Data Centre (NCDC), the National Oceanic and Atmospheric Administration (NOAA), and across a wide range of scientific journals and publications. In fact, the journals and publications have presented a grimmer view of what the climate science is really telling us, for far longer than acknowledged by the political and economic leaders of the world, policy makers, the IPCC, and any number of other global institutions including the World Bank, the IMF, the annual Davos Summits, the mainstream economic profession, and so on, all of who have downplayed the warnings from the concerned scientific communities. Sadly, even the IPCC between 2008 and 2013 was busy bending over backward to accommodate the special economic and political interests around the world by attempting to buy time for global capitalism—there is no other polite way of saying it—to the consternation of climate scientists around the world and those concerned with the fate and survival of humanity in the decades and centuries to follow. It is only in the last few years that the IPCC seems to have woken up again to its scientific and humanist responsibilities, in all probability too late to now stop the 2°C barrier from being breached. In that sense, the IPCC has already failed since one of its objectives was supposed to be to prevent the breaching of the 2°C "guardrail" erected by the climate scientists as a limit beyond which the heating of the atmosphere may become irreversible precisely because the positive feedbacks will take over the climate system. Future generations will certainly, on looking back, rue the lost opportunity of the time between 1992 and 2015 when significant action might have averted the approaching catastrophic events. All that is now water which has flowed under the bridge.

To an extent, environmentalists around the world were better equipped to understand what the climate science was saying because, thanks to their prior exposure to the science of ecology, they were in a position to connect the dots across disciplines, joining the abiotic and biotic interactions. However, in the fragmented environmental debate, it was always possible for certain countries to ignore environmental catastrophes in other parts of the world, for example, landlocked countries were not concerned about what happened to the oceans, desert countries could ignore deforestation in the Amazon basin, everybody wanted a finger in the pie of potential resources in Antarctica, etc. The emerging climate science neither permits such fragmentary and self-centric perceptions nor facile escape clauses based on dubious ill-informed arguments. In that sense, a major part of the environmental debate is a subset of the debate raging around the climate and not the other way around. There is no country on earth that will emerge unscathed from the ravages of global climate change because the nervous atmosphere touches every country on earth. Never before has the ground for international cooperation and action been stronger.

It is also possible to extend the interactions further, for example, by taking three system components at a time. For instance, heat accumulation in the oceans, which are absorbing 90 percent of the additional heat of the atmosphere, are causing more ferocious cyclones (an interactive effect between atmosphere, sea, and land). Major cyclones release energy calculated as equivalent to 100,000 Hiroshima-sized bombs or more, yet that is merely one-third of one day of global atmospheric heating. The destructive power of such cyclones is horrifying: a cyclone in Philippines a few years ago recorded gust speeds of 384 km/hour, at which speed the recording instrument broke down. The destructive power of wind is a cubic function of wind speed so that a 400 kilometer wind speed contains eight times more destructive power as compared to a 200 kilometer wind speed; however a 200 kilometer cyclonic speed is already a Category 5 cyclone, the highest category. Needless to add, such high cyclonic speeds have never been previously recorded.

Another three-component interaction that can be considered is between atmosphere-cryosphere-hydrosphere, wherein the melting of icecaps results in sea level rise; this is one of the slowest super-massive processes now already observed to be in operation but there is no way of predicting how much it will accelerate as global warming increases. Many major coastal cities around the world and several countries around the world (for example, Bangladesh and the Pacific Islands) face existential threat of a high order as do coastal areas all over the world, with an unknown timetable. Nevertheless, countries around the world are continuing to invest in expensive coastal infrastructure.

Yet another three-component interaction is between atmosphere-oceans-land in the form of increasing variance of precipitation on land. The increasing moisture content due to heating of the oceans/seas results in the dumping of larger volumes of rain/snow on the land; the higher wind speeds caused by increasing pressure differences between sea and land results in the variance, that is, alternation of floods and droughts over land areas. If one brings the fourth component, the biosphere, into this interaction effect, this will play havoc with surface vegetation over large areas of land. Living green matter is a delicate life form, it cannot survive continuous cycles of floods and droughts, it will simply die out, taking with it other life forms dependent on it in that biome.

One of the key understandings that the climate science has arrived at is that the atmosphere is in a dynamic balance with respect to emissions and absorption of CO_2 in the atmosphere. Every year, on a net basis, the earth system is able to absorb (ocean and land) about 3.1 to 3.2 billion tons of carbon per annum. As per IPCC 2007b (Vol. I, Tables TS 5 and SPM 2) the land absorbs net 0.9 billion tonnes of carbon per year while the oceans absorb about 2.2 billion tons of carbon per year. This represents an equilibrium limit up to which emissions into the atmosphere can be reabsorbed on an annual basis; any emissions exceeding this total quantity accumulates in the atmosphere and triggers global warming. The current level of emissions at a global level is between 9 and 10 billion tons of carbon per annum. The limit of 3.1 to 3.2 btCpa was probably breached around the middle of the last century. Since we have no record of the amount of explosives, fuel, etc. used in World War II, the climate record of CO_2 concentration in the atmosphere conveniently begins with measurements in the early fifties of the last century, by which time the atmospheric concentration levels had already started increasing and continue to do so. However, it needs to be noted that in the early fifties the developing countries including China and India had very small economies in terms of GDP and fossil fuel consumption so that the equilibrium limit had already been breached just by the emissions from the then-developed countries. Historically, most of the atmospheric carbon emissions have come from two fuels: coal, mostly used for the generation of electricity; and oil, mostly used for all forms of transportation, both passenger and freight. Natural gas as a fossil fuel was a late addition to the fossil fuel mix, probably in the sixties or seventies of the last century. Natural gas is a relatively cleaner fuel as compared to coal because part of the energy in LNG (liquefied natural gas) comes from the combustion of hydrogen in the fuel.

One of the easiest ways of understanding the root of the climate crisis is to look at the annual production statistics of the three major fossil fuels at a global level. These figures are compiled and published annually by the International Energy Agency (IEA) in their annual publication World Energy Outlook (WEO). Even with an elementary understanding of the underlying physics of combustion, it should not take more than fifteen minutes and a pocket calculator to compute the totality of global carbon emission into the atmosphere, on a reasonable assumption that the fossil fuel produced is consumed on a continuous basis and not stored. Such a computation will account for more than 90 percent of the carbon emissions at a global level.

The climate science is very clear in its implications: the carbon being mined from the earth as fossil fuels has to remain in the earth. If it finds its way into the atmosphere and if the amount of emissions crosses a threshold limit, the result will be long-term destabilization and possibly runaway heating of the atmosphere. Where the laws of nature are concerned, it does not matter whether a rich country or a poor country emits the carbon; it is indifferent as to whether a rich man or a poor man emits the carbon. The natural carbon cycle has operated over hundreds of millions of years to capture carbon from the atmosphere via photosynthesis and bury it in the earth in the form of fossil fuels. Promethean capitalism (and on a smaller scale, socialism) has dug it out and burned it over 250 years, using the atmosphere as a dumping ground for carbon dioxide, way above the absorption capacities of natural processes (sinks). In other words, unless total global carbon emissions are brought down to well below 3.1 to 3.2. btCpa and held at that low level for a long time, giving nature enough time to reabsorb the excess atmospheric carbon, global warming is inevitable. Essentially, global carbon emissions have to be brought down to zero if the climate is to be saved, and as rapidly as possible.

IPCC (2007a) has clearly indicated that a CO_2 concentration in the range of 350 to 400 ppm corresponds to a CO_2 equivalent concentration range of 445 to 490 ppm which in turn would result in a global equilibrium temperature range between 2.0 and 2.4 deg C (IPCC 2007b Vol. III). Volume I of the same report (*The Physical Basis*) also states:

> there is unanimous agreement among . . . climate models . . . that future climate change would reduce the efficiency of the earth system (land and ocean) to absorb anthropogenic CO_2. As a result, an increasingly larger fraction . . . would stay airborne in the atmosphere under a warmer climate. Atmospheric concentrations simulated by these models range between 730 and 1020 ppm by 2100. (IPCC 2007b)

All that remains to be noted is that atmospheric CO_2 concentration crossed 400 ppm in 2014–2015 and that emissions are continuing to rise. To the best of our knowledge, these tables and associated comments have not been subsequently retracted or disclaimed by IPCC, so they stand unrefuted. Since IPCC was mandated to try and hold global temperature rise below 2°C, by its own yardstick the IPCC has failed. We have earlier refrained from taking a harsh view of IPCC because IPCC has been under a lot of pressure from climate-denier lobbies which have been extremely well funded by the oil and coal lobbies, particularly from the United States.

Yet, IPCC cannot rescind the scientific findings, no matter the pressure, because that would be a betrayal both of good science and of human futures.

Many observers not tracking the deeper scientific debates revolving around climate change may feel that the projections about the future are alarmist. They would be wrong. If anything, most climate scientists have soberly understated the most dire implications of their science which is that global warming caused by anthropogenic emission will affect the global climate for centuries or millennia to come, that the climate system is far too large and uncertain for humanity to control its titanic forces, and that irreversible climate change (when the larger positive feedbacks kick in) may well destroy most life on earth, a sixth mass extinction.

Part of the hesitation on the part of the scientific community—their reserve—stems from the many unknowns within the science itself, for example, the magnitude of the climate sensitivity, the timings at which various positive feedbacks get activated, model uncertainties, time lags between emissions and temperature rise, and so on. For example, the common belief within the scientific community was that there was a time lag of thirty-five to forty years between a given level of CO_2 atmospheric concentration and the corresponding equilibrium temperature being attained. Recent findings however seem to indicate that the time lag is closer to ten years (Ricke and Caldeira 2014). This means that temperature rises will occur much faster than earlier expected and consequently, the trigger of positive feedbacks may occur much earlier—in plain words, we are living in a fool's paradise. We are much closer to the desertification of the earth, the destruction of forests and habitats, the disappearance of glaciers, an increase of heat waves and cold waves, increasing ferocity of floods, storms, and cyclones, disappearance of polar ice, and the many other consequences to follow. This will be the true inheritance bequeathed by us to future generations, and not an increasingly "richer" world as is being promised. Yet few within this generation seem to have the courage to say so. That is the tragedy of the present.

THE INTERSECTION OF THE CLIMATE CRISIS AND THE GLOBAL POLITICAL ECONOMY

The Rio Conference 1992 marked the establishment of the UN Framework Convention on Climate Change (UNFCCC) and under its aegis, the IPCC. Six years down the line, one of the outcomes of the Rio Conference was the Kyoto Protocol which required the developed capitalist countries (politely referred to as Annex I countries) to reduce their atmospheric carbon emissions by 5.2 percent by 2012 with the base year being 1990. The actuality of what has happened may be judged from the following: In 2013, the emissions reduction for the first Kyoto Protocol period was computed at 16 percent, and this was hailed as a "runaway success." The KP required computation of territorial emissions, that is, all emissions from a territory. This is at variance with consumption emissions of a territory, which is a true reflector of the emissions of a territory. In consumption emissions, embedded carbon emissions of imports are added and those from exports are subtracted. The consumption emissions of North nations for the first KP period increased by 23 percent whereas territorial emissions decreased by 16 percent. Outsourcing of goods and services consumed by North nations to South nations, particularly China and India . . . shifted the crediting of a significant amount of trade emissions from North to South nations (Dhara 2019). In other words, the Kyoto Protocol failed in the first commitment period.

Not only has the Kyoto Protocol failed in the first commitment period 1998–2012, it is headed for failure in the second commitment period 2012–2020 in another very significant dimension. During the first commitment period, Northern countries were legally committed to collectively contribute USD 30 billion per year into a Green Environment Fund (GEF) to be administered by UNEP and used by Southern countries for purposes related to climate mitigation and greening the environment. The second commitment period required the contribution to be raised to USD 100 billion per annum. There has been a collective failure by the capitalist countries to honor their treaty obligations—they have failed to make the necessary contributions.

This stingy behavior of the capitalist countries—there is no need for polite words—needs to be seen in the context not only of the combined GDP of the United States and Europe (around USD 43 trillion in 2018) but also the kind of wealth in private capitalist financial corporations centered in the United States and Europe. The world's top seventeen asset management firms have assets under management totaling USD 41.4 trillion. Each of the firms that exceeds USD 1 trillion as assets under management are referred to as giants. Another three similar firms, each with assets under management ranging between USD 1.0 and 1.2 trillion, are referred to as new giants and their combined assets total another USD 3.3 trillion. The next nine firms, each with assets between USD 800 billion to 1 trillion, have combined assets of USD 8.15 trillion. Together, these twenty-nine firms have assets under management nearing USD 53 trillion (Phillips 2018, pp. 37–52). This is part of the combined wealth of a transnational capitalist class and draws what could be considered "rentier incomes" from the worldwide capitalist (and no doubt, socialist) economy. The situation is similar to the eighteenth and nineteenth centuries in Europe where a small landowning feudal class appropriated a vast share of income based on ownership of land, the difference being that this time it is a capitalist class which is worldwide. In other words, it is not as if there is not enough money in the Northern countries to pay a lousy 30 billion or 100 billion per year; it is that the money is in private hands, fully protected by state power.

In the course of the French Revolution, before losing his head, Louis XIV had exclaimed: "Après moi, le deluge!". Surveying the working of the transnational capitalist class today, he might have remarked, "Après vous, le desert, le heat and cold waves, le sea level rise from dying seas, le droughts et le deluges, le cyclones et le tornadoes, le forest fires, le extinction de species, et le radioactive waste."

It will probably require some kind of a revolution to get the money back into the public domain and use it for saving the planetary climate system.

This is the core of the climate crisis which we have chosen to label as certain death. Viewed from the perspective of the global political economy, the above picture is deeply disgusting. Those who are naïve enough to believe that the holders of this wealth (or capital) will willingly part with it for planetary purposes are welcome to continue with their beliefs. It is for those who find the situation deeply disturbing, since it has been done in their name as citizens of democratic nations, that the need to find a way out is being addressed, as a call toward the greening of socialism.

The case for a reversal of global monetary flows has been made by Bellamy Foster in some detail, as an ecological debt: it is better that this case is argued within the capitalist countries by those citizens of the countries who are now convinced that the present course is politically no longer acceptable. Such a discourse emerging from Southern countries has hitherto fallen on the deaf ears of the transnational capitalist class. Verily hath it been said: there are none so blind as those who will not see.

Before bringing this section on the climate crisis to a close, an honorable mention needs to be made of the role of the United States. Post 1998, it signed on to the Kyoto Protocol during the Clinton administration. Under the Bush administration, it withdrew from the Kyoto Protocol, citing the emissions from China and India for whom Kyoto was not binding. At the Copenhagen Summit in 2009, which was a failure, the United States under the Obama administration tried to grab victory from the jaws of defeat through some last minute political grandstanding. Under the Trump administration, it walked out of the Paris Agreement. It is quite clear that the US political system is hostage to business interests spearheaded by the fossil fuel industries led by oil and coal and puts its business interests above those of a climate system headed for destruction. The United States represents the core of the global capitalist system and its actions have amply demonstrated its priorities. The rest of the world is not going to be bothered about how each individual state in the United States responds to the climate crisis through its policies. It is for US citizens to bring their administrations into line with reality. As regards China, IEA figures for coal consumption in China will indicate that in the decade 2005–2015, China decreased its coal consumption by over 1 billion tons per annum, which is larger than the entire annual coal output of the United States at around 900 million tons per year. China is a middle-income country on a rapid path of economic growth; such an endeavor risked derailing its entire growth which pulled hundreds of millions out of historic poverty, yet China took the risk. This is an inconvenient truth not to the liking of the capitalist-controlled press and media in the Northern countries which would like to paint China as the new threat to the Free World. Another data-based truth about recent performance: China added 44.3 gigawatts of solar power in 2018, the United States added 10.6 gigawatts and India added 8.3 gigawatts in 2018 (Kabeer 2019). Though official figures have not been released, Japan installed 7 gigawatts and Germany 3 gigawatts with the entire EU at 8 gigawatts. Note that the Chinese annual installation is much more than the combined annual installation of the United States, the

EU, and Japan—so now we know who is straining to save the world's climate. Incidentally, China's installation in 2018 was significantly lower than in 2017 (53 gigawatts), so it is not a flash in the pan. It is the difference between posturing and action.

Chapter Five

Critiques of Models of Political Economy

In the previous three chapters, we have tried to establish the links between each of the three crises (nuclear arms race, the environmental crisis, and the climate crisis) and the overall structure of the political economy of capitalism. In this chapter, we shall quickly overview four of the main models of political economy that exist or are emerging around the world: Hayekism, Keynesianism, state socialism, and emergent green Keynesianism. Before proceeding further, a brief cautionary note about the use of the term "model" henceforth: it is not the same meaning as used earlier in the discussion on climate change. There, the term "model" refers to the scientific-mathematical models used in the climate science to explore and project variables and their relationship and interaction as understood in any of the disciplines of the natural sciences—physical or biological. Here, the term "model" is specifically used to sum up a historical process within political economy related to the circulation, accumulation, or distribution of the social surplus within an economy.

The circulation of surplus within a capitalist economy has been a matter of considerable interest in Marxist debates on political economy because Marx (in *Das Kapital*) had posited that the exploitative relationship of capital with labor would eventually result in a collapse in demand which would lead to a systemic crisis of capitalism, with the distinct possibility of the overthrow of capitalism by the working classes within the affected country. The Great Depression starting in the United States in 1929 was one such moment in the history of capitalism, which not only resulted in a global crisis but was a principal causative factor which precipitated World War II. There have been several economic depressions, recessions, bubble bursts, and meltdowns in nearly a century of economic activity undertaken by capitalism since then, the latest and one of the most severe manifestations being the financial crisis of 2008, again stemming from the United States, which threw the entire capitalist system into panic on a global scale. It required coordinated emergency global action to prevent a collapse of the worldwide financial system including banks, insurance companies, pension funds, stock exchanges, currency exchange rates, international payments and settlements, and so on.

Yet, the capitalist system has survived these periodic crises and Bellamy Foster has been courageous in admitting

> Marx and Engels clearly thought that the duration of capitalism would be much shorter than earlier modes of production, brought to a relatively rapid end by the intensity of its contradictions and by the actions of the proletariat—the gravediggers of the system. (Foster 2009, p. 231)

The other school of thought, O'Connor and Kovel et al., have argued that there is another "external" contradiction between the capitalist system and nature which can potentially drive the capitalist system toward collapse. Capitalism, therefore, is a system caught between two modes of potential collapse—an internal one due to a crisis of demand and an external one due to a crisis arising from nature. Ergo, if climate change is an external crisis of enormous magnitude, it may well result in the collapse of capitalism. This is the fear that haunts the global capitalist classes, that the system in crisis may be nudged toward collapse with a little help from the gravediggers, internal and external. That the capitalist system, from 2008, has been under enormous stress seems apparent, yet it has not collapsed; it continues to hold a major part of humanity in thrall by assiduously propagating the belief that there is no alternative to capitalism. A look at these four models of political economy, in brief, may serve to throw light on the above assertion and the possible emergence of alternatives. For a somewhat deeper examination of these models, the reader may refer to my earlier writing in the journal *Capitalism Nature Socialism* (Ghotge 2018).

MODEL 1: HAYEKISM (NEO-LIBERALISM)

In a 2014 lecture, Wolfgang Streeck traced the roots of the present phase of neo-liberalism back to its ideological roots in the theories of Friedrich Hayek and Milton Friedman, developed in the thirties and forties of the last century (Streeck 2014). These theories were revived in the mid-eighties particularly in the UK and the United States and gathered strength in the early nineties with the collapse of the Soviet Union and the East European states, which allowed global capitalism to expand unfettered across the world. Although a range of critical authors have referred to these revived theories by a variety of names—New International Economic Order, Reagan–Thatcher reforms or just reforms, Washington Consensus, Monetarism (the Chicago School), neo-conservatism and more politely neo-liberalism—yet at the core they emanate from Hayek, hence we refer to this model as Hayekism. At its core lies the belief that a free market system, created and protected by state power through reforms, must enable the flow of the social surplus from the bottom of the social pyramid to the top and anything that impedes such a flow of the social surplus is not in the interests of society, be it protections granted to labor or the environment, etc. David Harvey has surveyed the range of neoliberalism's global reach and impact in particular from the international perspective on labor (Harvey 2007). Hayekism, with its emphasis on the unimpeded extraction of social surplus, flies in the face of Marx's prediction about a capitalist crisis of demand created by its own internal mechanism which halts the circulation of capital. The vision of unbridled capitalism accords to the state the role of creating and implementing policies aimed at the removal of all barriers to the accumulation of capital at the top of the pyramid. The removal of prudential barriers between the various circuits of capital—the stock markets, the banks, insurance, pension funds, etc.—has been a major causative institutional factor in the financial meltdown of 2008. During this period, sarcastically referred to as the socialism of the rich, the real crises referred to in the previous chapters have been ignored as global capitalism in crisis has desperately striven to buy time to stabilize itself notwithstanding the growing perception around the world that capitalism as a system has no answers to the real existence-threatening crises engulfing the globe. Its only answer seems to be a completely self-serving answer: the continued growth of capitalism as a global system to which no alternative is possible.

MODEL 2: KEYNESIANISM

Keynes, also in the thirties and forties, developed an alternative theoretical formulation to deal with the capitalist crisis. Since the crisis emanated from the collapse in demand, Keynes postulated the revival of demand through concerted state action to revive demand. State action could regenerate employment through state-funded public works such as infrastructure building. This employment would put incomes in the hands of the formerly unemployed workers who would then spend the money to buy the unsold goods on the market, that is, generate demand and kick-start the circulation of commodities which had come to a halt as a consequence of the crisis, the resultant unemployment, etc. The state would fund its political intervention by printing additional money; this would result in inflation which was preferable to the recession/depression that was caused by the crisis. In short, an inflationary economy was preferable to an economy in recession which could rapidly spiral downward resulting in a cyclic process of collapsing demand, business closures, unemployment leading further to collapsing demand, and so on. Keynes was not too concerned with the kind of economic activity that was to be generated through state intervention—in his famous words, "let them dig holes in the ground"—so long as the overall economy was revived. Since getting the economy moving again would take time, Keynes also proposed a range of state-funded welfare measures to help the population in deep distress.

There is no need to doubt Keynes' humane intentions—he had beheld the horrendous consequences of WWI, the slaughter and the economic destruction—and tied it to the economic foundations of the crisis. In his desire that such a cycle should not recur, Keynes also proposed a triad of global economic institutions—the Bretton Woods institutions—whose function was to provide the necessary wherewithal to keep the economic cycle in motion. The three institutions he proposed were: the International Bank for Reconstruction and Development (the IBRD later popularly dubbed the World Bank), the International Monetary Fund (the IMF), and the International Trade Organization later renamed the World Trade Organization (WTO). The first two were established immediately after WWII while the third was established only in the nineties. Between these three institutions, a foundation was laid for the globalization of the capitalist economy by opening out the channels for the flow of all three forms of capital—financial, industrial, and commercial—as well as the flow of commodities between countries.

At its core, the Keynesian paradigm represented the re-routing of a part of the social surplus from the top to the bottom of the social pyramid in the capitalist societies. This enabled them to reach full employment-full output in their economies which grew rapidly after the reconstruction following the destruction of WWII. This period of nearly a quarter century (1950–1974/5) is often referred to as the Golden Years of capitalism and Hobsbawm has devoted an entire chapter to discussing the period (Hobsbawm 1994). As far as the capitalist countries were concerned, this re-routing of surplus did work, bringing with it an age of stability, consumer prosperity through rising incomes, and a level of mass welfare centered on education, health, transfer payments toward social security, and so on. Yet the picture was nowhere near as rosy where the rest of the world economy was concerned. Despite Keynes' humane design, the international financial institutions under the control of the capitalist countries hardly lived up to their promises and the expectations they aroused.

Keynesianism failed in many unforeseen and unexpected ways, not because the intentions were questionable but because of the actual working of the capitalist system. It resulted in military Keynesianism, a global arms race based on huge budgetary allocations for defense purposes, the active creation of a military-industrial complex. It was at best a meta-stable solution which ensured prosperity provided economic growth continued to provide the social surplus that could be redistributed to ensure social peace; rather like maintaining the balance on a bicycle only so long as the bicycle keeps moving, if the bicycle stops the rider will fall off. This is the root cause why consumption in the developed countries cannot be reduced through policy measures because it risks precipitating a systemic crisis. Moreover, Keynesianism also proved to be reversible; once neo-liberalism was back in the saddle in the eighties and nineties, the social benefits deriving from the Keynesian paradigm were progressively dispensed with by the new political dispensations that came to power in capitalist countries. Lastly, the Keynesian paradigm did not really change the class relations under capitalism—the first contradiction—in any meaningful way; the prosperity of a couple of decades served to blunt the sharp edges of the underlying contradiction. Moreover, Keynesianism preserved intact the exploitative machinery of capitalism operating internationally across the countries of the Third World, whether in large parts of Asia, Africa, or Latin America; in fact, it provided the institutional tools and frameworks which were manipulated by the capitalist system during the following period of neo-liberalism and the global expansion of capitalism.

There is a lesson of history, harsh perhaps, for those willing to understand it, namely, that a return to Keynesianism without significant resolution of the underlying class contradictions cannot now provide the answers to the real crises confronting the twenty-first century. Keynesianism was an interregnum that served to conceal the iron hand beneath the velvet glove, before the ruling classes returned to business as usual and politics as usual. Mainstream economics would like us to believe that the only choice is between Hayekism and Keynesianism, with these alternating in turns. Mainstream economics is akin to a priesthood: just as the priesthood of times past wanted us to believe that God created the world ex-nihilo, whereas it was the priesthood that created God ex-nihilo, so too our bourgeois economists would have us believe that the creation of money ex-nihilo by the state, veiled behind the mists of misplaced reverence and awe, is the ultimate answer to our current crises. Money is a man-made variable, the real crises are mostly generated by natural variables now out of control.

The monetization of damages caused to and by nature is a part of the cognitive fraud that is sought to be practiced by the adherents of mainstream economics who cannot seem to get it into their heads that the forces of nature are far larger than anything that mankind has the capacity to withstand. Mainstream economics is hubris on stilts. Moreover, the creation of money also seems to be out of control and has resulted in massive public and private debt, huge reservoirs of private wealth outside the control of any legal or regulatory powers and a system of over-production and over-consumption that now poses a threat to all of nature and in consequence, to all of humanity.

MODEL 3: STATE SOCIALISM

Just as Keynesianism represented a re-routing of social surplus from the top to the bottom, the state socialist model also re-routes surpluses to the bottom of the social pyramid. In theory, there is supposed to be no class of private capitalists, the state is the sole capitalist leading some observers (starting with Trotsky) to name this model as state capitalism. However, there are huge differences between a capitalist economy and a socialist economy in so far as much of the productive apparatus in the latter consists of state-owned enterprises, co-operatives, worker-owned enterprises, and a comparatively smaller number of private enterprises. A major part of the commanding heights of the economy are generally under state ownership or control, for example, the energy sectors, banking, railways, mining and minerals, pensions and insurance, heavy infrastructure building, the water sector, and so on; private sector activities range from agriculture and livestock sectors to trade, hospitality, stock exchanges, and the new technology sectors—communications, computers, and so on. China has been the most prominent example of a model 3 economy. During the post-Maoist period, that is, the last forty years, it has exhibited the most remarkable sustained economic growth witnessed in world history. At an average growth rate of 10 percent per annum, an economy doubles every seven years so that over a forty-year period the size of the economy would be well over thirty-two times what it was at the start of the process. China has managed this transition over this period, pulling well over 500 to 700 million of its poorest population to a level where it now ranks as a middle-income country. This economic performance far overshadows anything that the capitalist countries managed in terms of the removal of absolute poverty over two centuries of global dominance. China is now the second largest economy in the world and fast catching up with the United States in terms of sheer size though it lags behind in technology, higher education and research, quality of life for a majority of the population, and so on. However, by any yardstick, it has left historical poverty far behind.

China started on its present path with the ambition of becoming "the workshop of the world" rather like Britain had been in the eighteenth and nineteenth centuries. This ambition it has achieved but at a price—to the environment, to the climate system, and possibly also to its own people. Some idea of the environmental costs it has incurred can be found in Watts (Watts 2010). On the CO_2 emissions front, with national accounting of emissions, China is now the largest emitter in the world but, as indicated earlier, if consumption emissions are considered (that is, national emissions plus imports minus exports) then the picture is quite different since exports to developed countries form a significant part of Chinese production. These exports from China would get added to the emissions from western countries (as their imports) revealing the more correct global picture. In this sense, China has been caught in a trap of its own making, with huge overcapacities in energy intensive sectors such as steel, cement, fertilizers, and coal-based electricity, all of which are huge carbon emitters. More worrying are the future directions of Chinese growth, at least from the perspective of climate change. Two in particular need to be noted—Xiongan and One Belt One Road (OBOR), both having global implications.

Xiongan New Area is a huge greenfield city project being developed by the government of China, a city to house over 130 million residents. To put matters in perspective, the largest cities in the world today house around twenty million residents each, so that Xiongan would be a monster of a city. Put another way, it would house 10 percent of China's current population. The city, work on which commenced in 2017, is planned as a futuristic "socialist" city, a prestigious project supported by China's top leadership. It is in the middle of a triangle formed by Beijing, Tianjin, and Hebei and the official briefing states that as much as USD 580 billion could be spent on constructing it over the next two decades. The city will have all the hallmarks of a "smart city"—ultra fast transportation networks, water and energy conservation technologies, high speed communications technologies—and is intended to act as a hub for future cutting-edge technology industries as well as to relocate many of the non-governance elements of Beijing city. It is part of a strategic vision to move about 400 million of China's rural population into urban surroundings and to act as a model for other such urban agglomerations in the future. In part, it is envisioned as China's Silicon Valley for attracting talent to futuristic industries. Currently, the official build-up surrounding the project resembles a high-profile real estate developer in the capitalist world marketing prime property in the future, promising all things to all people. It is too early to pronounce judgment on any gap between promises and performance, but the scale is certainly ambitious, on a scale never seen before. Nothing in the current plans indicates whether this is a response to future sea level rise due to climate change, yet it is a strategic gamble, being undertaken at a time when there is a huge real estate bubble in urban housing in China, with hundreds of "ghost townships" lying unoccupied across China with millions of apartments, thousands of malls and shopping centers, and vast urban infrastructure and urban transportation systems reportedly lying idle. This unproductive investment is apparently already affecting the balance sheets of its banks and financial institutions, coming as it does on top of huge overcapacities in steel, cement, fertilizers, and power already mentioned earlier. More to the point, the alienation of man from nature will be nearly complete in such an undertaking, posing fundamental questions about a green socialist society of the future. Nor are there any indications about the possibility of the state "withering away" even in the remote future. If the former Soviet era was fascinated with huge projects, this undertaking dwarfs them completely.

The other Chinese strategic undertaking, on an even more ambitious scale, is the One Belt One Road initiative, about which much has already appeared in the media. In essence, it consists of two core elements: a maritime component connecting China with Africa and beyond that, with Europe; and a land component connecting China via Central Asia and Russia with Europe. The land component will also be extended via Southeast Asia to Australia by high speed rail links and infrastructure. China has also made huge investments in securing its future energy supplies (oil and gas) from Russia, thus competing directly with Western Europe. To be fair to China, this is obviously a response to future encirclement and interference with its strategic supply routes by the western powers; it is also an answer to its excess production capacities, its dependence on prosperous foreign markets, its need to invest capital abroad, and its need to employ its huge manpower in large projects which it will fund. The patterns are distressingly familiar to the earlier epoch of western colonialism, except that they are being undertaking by a country with a Communist Party in control. One possibility for the future is that it may loosen the capitalist hold over many countries around the world by offering an alternative avenue for development even if the lines of development do not seem to be very different from those offered by capitalism. The other possibility is the emergence of "colonialism with Chinese characteristics." Again, to be fair to the Chinese, there have been no overtly aggressive actions on their part in the last four decades—if anything, they have been more sinned against than sinners—but the balance of power is rapidly shifting, not merely in East and Southeast Asia but around the world. There is cause for worry because the discourse in the English-language western media is consistently critical of China indicating that the Cold War mindset has not disappeared.

Both Xiongan and OBOR are projects for the future but Xiongan, prima facie, hardly seems the kind of project that would transcend the metabolic rift that formed such an important part of Marx's theorization. Quite the contrary, it seems to mark the complete domination of the city over the countryside. Nor is the substitution of ersatz nature interspersed into the city likely to overcome the alienation of man from nature, or lead to a great transition from capitalism to socialism. OBOR may be a response to the productive capacities of China, including having developed a massive working class as a result of having become the workshop of the world, but putting that working class to work on projects all around the world can hardly be counted as furthering the cause of socialism in China. In any case, both these projects represent a strategic gamble for the future and a transition toward socialism in China may well hinge upon the success of both; socialism, after all, is about people and not about the economy which is a means to a political end.

MODEL 4: GREEN KEYNESIANISM

Green Keynesianism is an emergent model, a work-in-progress, a re-invented form of capitalism under construction. At its core, it consists of replacing the black (that is, carbon-intensive) parts of the capitalist economy with green counterparts, whether in energy, materials, commodities, production processes, or living and working spaces including residences, factories, offices, or even entire "smart" cities. Thus, fossil fuels will be replaced with green/renewable energy systems; petroleum-based transportation with electricity or hydrogen-fueled transportation systems; petroleum-based materials (plastics, polymers, petrochemicals) with materials sourced from biological sources; other materials being produced through alternative processes not involving carbon (for example, steel and cement); material substitution using newly developed materials; food sourced from organic processes or genetically modified processes; transportation demand being substituted by information transfer; entire new ranges of "green" or "smart" commodities being produced in response to the crises of climate change and the environment. Entire cities will be redesigned to make them green or smart or whatever label that sells, including all the spaces within the cities be they residential, commercial, office, factory, recreational, educational, tourist, and so on. Even warfare will not remain unaffected with smart bombs, smart missiles, precision weapons, surgical strikes, possibly entire armies of robots fighting each other, controlled by GIS systems mounted on satellites. This "revolution in military affairs" (RMA) has already started. As mathematicians Davis and Hersh presciently noted:

> One began to hear it being said that WWI was the chemists' war, WWII was the physicists' war, WWIII (may it never come) will be the mathematicians' war. (Davis and Hersh 1981, p. 93)

This brave new world of green Keynesianism that is being constructed has many names: a Green New Deal (Barbier 2010), a Post-Carbon Economy, a New Energy Economy, a Green Economy, and so on. Elements of its emergent design can be seen in the writings of Weizsacker et al. (Weizsacker et al. 2009) and earlier in Weizsacker, Lovins, and Lovins (Weizsacker, Lovins, and Lovins 1998). These writings indicate the technological possibilities across diverse fields of production wherein the current levels of output could be maintained while reducing the inputs by factors of four or five, hence the titles of the books.

What is common in this approach, with the earlier vision of Keynesianism is that the underlying contradiction within capitalism—the first contradiction—is left unresolved. As with earlier Keynesianism, its military version remains a distinct possibility despite the earlier experience with Keynesianism during the Cold War. When the Cold War ended, around 1989–1990, for those who care to remember, the hope had been held out of a "peace dividend" which never materialized. There are, therefore, no grounds to believe that a renewed set of promises will be fulfilled. All that green Keynesianism will achieve is to provide capitalism with more time to set its own reign in order by diverting the social surpluses into a newly restructured capitalist economy. The promises of establishing a Green Environment Fund by contributing USD 30 billion per year have proven hollow in practice so that the poorer countries of the world will largely have to fend for themselves in the face of destructive climate change not of their making. There is already enough technology available to have begun a serious transition away from the black economy—efforts such as a much greater emphasis on public transportation, coherent public policies for renewable energy, stopping the privatization of water resources, curbing wasteful and frivolous consumption and the generation of mountains of urban waste are all policies that could have been pursued in the last quarter of a century and might have made a critical difference to the future. Instead, the emphasis has been on undoing the damage caused by the instability of the capitalist system and its man-made monetary crisis—the dot-com bubble, the East Asian crisis, the great meltdown in 2008, and a series of unnecessary wars—in Iraq, in Afghanistan, in Libya, in Syria, and so on, all making global cooperation that much more difficult. Now, even the adolescents are saying: if the system cannot change, then change the system; and the children are telling the grown-ups to behave like adults, a tragi-comic reversal of roles.

Chapter Six

The Greening of Socialism

In the foregoing analysis of four models of circulation of the social surplus, we do not find convincing answers to the global existential crises confronting human society in the twenty-first century nor to bringing nature explicitly into the framework of analysis. The greening of socialism, as proposed in a structured "thought experiment" as model 5, outlined below, is a preliminary attempt to think through the future. It is neither a blueprint nor a plan, nevertheless it is an attempt to learn from past failures on a global scale while keeping the future open. If the reading and comprehension of history serves any purpose, it is to prevent repeating the mistakes of the past and to resolutely concentrate attention on the future. Capitalism has continuously been in denial of the reality of the three existence-threatening crises previously outlined, substituting instead its obsession with its own fictitious commodities—land, labor, and money—along with other newly invented fictitious commodities conjured out of thin air (ex-nihilo) through the creation of fictitious property rights on a global scale. This system, based on multiple fictions, is a house of cards which continues to totter every time the winds blow. Yet capitalism has refused to collapse under the weight of its own inner contradictions much to the disappointment of revolutionaries of various hues and the relief of its diehard apologists and mythmakers. In the twenty-first century, it is becoming increasingly evident that the internal contradictions have now been joined by a set of external contradictions emanating from nature and mankind's relationship of seeking to dominate nature, so widespread under capitalism. Since capitalism refuses to oblige by collapsing, perhaps a little helping hand from all the other classes, both in the center and the periphery, seems called for, yet this is merely a beginning for a great transition toward a socialism that brings together the concerns for the survival of both human society and nature, taken together; this beginning of a great transition can no longer be postponed.

The model of green socialism reallocates the social surplus from the top of society to the bottom but also further back, to nature. The distinctive feature of green socialism, which distinguishes it from the other three models of capitalism (Hayekist unbridled market capitalism, Keynesian welfare capitalism, and green Keynesianism) as well as from state socialism is also its core concern:

> What Fourier referred to as the "third and last emancipatory phase of history"—the first having made serfs out of slaves, the second wage earners out of serfs—which will transform the proletariat into free men by eliminating the commodity character of labor, ending wage slavery, and bringing the commercial, industrial and financial institutions under democratic control. (Chomsky 1970/2005, pp. 18–19)

This is a political objective, enveloping both society and nature, which cannot be achieved without overcoming capitalism itself.

Green socialism is a paradigmatic historical alternative to capitalism which enhances nature's productivity, reverses the destruction of nature, reduces the quantitative material and energy throughput of the human economy, reduces waste, and reduces entropic cycles of disorder while stabilizing employment and incomes through nature-directed activities/programs/projects paid for through reallocation of social surplus. During the period of transition, economic activity can occur through any of three regulated organizational forms: state-owned enterprises, producer-owned cooperatives, and private enterprises; and newer forms can also be developed, without losing sight of the political objective of the global emancipation of labor.

The green socialist alternative can register high growth, though the metrics of growth would be entirely different as compared to capitalism and would be both qualitative as well as quantitative. They would also be different between developed countries and developing countries—in the former, they would be far more centered on building up people through cooperative activities (economic, educational, and cultural) in a reorientation away from the excesses of both competition and hyper-individualism. In the developing countries, many of which still require modern infrastructure based on material growth, the green socialist paradigm would work out pathways which are neither destructive of nature or human societies. Some indications are given in the paragraphs that follow, in the form of building blocks; yet, commodity production and consumption would no longer be the centerpiece of economic life unlike under the most advanced form of welfare capitalism being conceived today, that is, green Keynesianism.

Earlier societies were aware of nature's contribution to human welfare and were willing to accept that some part of surplus, perhaps as collective labor, was needed to return to nature its due share. It is only under capitalism that nature has been viewed as a source for limitless expropriation and a sink for limitless waste. Green socialism can no longer continue with such values, as we have ample and mounting evidence, because they could result in the complete collapse both of nature and of human society.

With the money appropriated through the reallocation of military budgets and the end of military Keynesianism, the following priorities toward nature have to be considered: (1) returning carbon to the soil and to the ocean deeps through phytoplankton, (2) reforestation with natural species in appropriate habitats, (3) cleanup of all fresh water bodies—streams, rivers, lakes and where necessary, even polluted aquifers, (4) cleanup of seas and oceans including restoration of marine breeding areas, mangroves, and coastal breeding grounds, (5) restoration of desert and semi-arid areas and species, (6) stabilization of mountain slopes through forest restoration, and (7) sequestration and safe disposal of nuclear wastes as a special category. This reallocation of global surplus is a better alternative than the fatuous Keynesian exhortation to "dig holes in the ground." There is a profound need to help nature heal itself, if only from our anthropocentric concern for human survival on earth.

While a steady-state economy is probably impossible on a global scale, during the phase of transition, except as an idealistic construct, it seems possible to have a highly developed human society living well within the bounds of the various cycles of nature, constructing which is well within our collective capabilities. However, since collective capabilities are neither recognized nor respected within the competitive framework imposed on societies living under capitalism, the tasks required to be undertaken will have to perforce be conceptualized outside that framework. The genuinely important human needs: good quality public health, high quality education, an adequacy of modern infrastructure, nature as common property and common global heritage, smaller and less alienated communities, free technological knowledge for meeting future needs, an end to the hoarding of nature as private property, less drudge work and more free time, all have to be considered. Revitalizing nature and recreating human community have to go hand in hand.

MODEL 5: THE BUILDING BLOCKS OF A GREEN SOCIALIST POLITICAL ECONOMY

As indicated above, model 5 is a "thought experiment" to push forward the outlines of an emergent process that appears to be coming into being on a global scale. From all indications, we have to now start preparing to save as much as possible of human society and of nature from the collapse toward which it seems headed, both in this century and for centuries to follow. This exigency requires us to move well beyond general principles and long-term objectives toward an alternative framing of long-term policies and programs resulting in a qualitatively different world of the future. We put these, very briefly, as the building blocks for a green socialist political economy. Each building block may be viewed as a skeletal flexible set of internal relations, yet the relationship between the building blocks is important to achieving the desired long-term objectives.

Objective 1

A global limit on war production and defense budgets, with a delegalizing of nuclear weapons with no exceptions, under the Vienna Convention banning chemical and biological weapons. The entire nuclear industry to be brought under the ban, both since civil and military uses of nuclear materials are co-dependent and we have no moral right to leave nuclear wastes for future generations to clean up. The reduction of military budgets will have to be brought about in each country through the democratic pressure exerted by its citizens, and will result in enormous budgetary resources—the means—to fund the future transition. It is logically completely absurd to expect global cooperation for saving both humanity and nature while countries threaten each other continuously with nuclear weapons, missiles, stealth weapons, cyber warfare, and the whole military infrastructure necessary for instant mutual annihilation. It is only the citizens, particularly in the developed countries, who can decisively change this scenario through democratic processes. The importance of achieving this objective cannot be over-emphasized; failure to achieve this objective may possibly result in the failure of all other transition objectives. A start can be made immediately toward a total ban on nuclear weapons since 121 countries in the UN have moved for such a proposal at the global level.

Objective 2

A complete overhaul of the entire taxation policy including subsidies to corporations using fossil energy, personal taxation, a huge variety of "incentives" to promote growth, wealth taxes and estate duties, and so on, with a corresponding relaxation of personal taxes on those with lower incomes and wealth. If the owners and CEOs of large corporations pay themselves ten times or more of what they are worth on the market, it is quite in order for legislation to take away 90 percent of their incomes through taxation, thereby restoring parity.

Both of the above measures will be hugely resented therefore resisted by the dominant classes and their mouthpieces in the media. These changes have to be legislatively enacted so that there is no arbitrariness in the procedure involved, which will require democratically ensuring that the powerful are given a clear signal that their dominance is at an end. Moreover, since these will generate the necessary revenues for the other parts of an agenda of transition, they may be seen as providing the means and not as ends in themselves. If these measures cannot be successfully enacted, there is no way of achieving a transition.

Objective 3

Curbing the production of money "ex-nihilo," in all its forms, both by the state and private corporations (though gearing ratios) leaving a huge burden of public debt (national debt) while private wealth beyond any regulation grows out of all proportion or control. Many curbs on the financial sector which have led to an international "casino economy" are now overdue; all self-regulation under capitalist rule has failed in this regard, leading to a huge number of global tax havens where money can be conveniently hidden or laundered. This will also be fiercely resisted.

Writing in 1970, Chomsky said:

> Its essential that a powerful revolutionary movement exist in the US if there is to be any reasonable possibilities for democratic social change of a radical sort anywhere in the capitalist world. . . . At the very least it requires that domestic pressures impede the great centers of world imperialism from counter revolutionary intervention. (Chomsky 1970/2005, pp. 34–35)

That moment has now arrived.

Objective 4

A wide range of pro-nature activities to be publicly funded with particular emphasis on the global restructuring of farming. Capitalist society has hitherto refused to recognize that farming is potentially carbon-negative and needs to be revalorized. Just as carbon emissions into the atmosphere need to be taxed, carbon absorption from the atmosphere needs to be globally subsidized. This single measure would significantly usher in a process of reducing global inequality, stabilizing demand in many developing and developed countries, reduce rural-urban distress migration, restore health back to the soil through organic agriculture (that is, basically returning carbon back to the soil, as nature does), and produce better quality and more healthy food for humans and livestock including species in the wild. Industrial agriculture in the last two centuries has been one of the most destructive human practices on a global scale, resulting in long-term destruction of soils, water, and biodiversity along with the destruction of peasant societies. The latest and first UN report on global biodiversity, released on May 6, 2019, the IPBES 2019 (Intergovernmental Science—Policy Platform on Biodiversity and Ecosystem Services) estimates that of the 8 million animal and plant species on earth, 5.5 million are insect species; we have simply no idea what would happen if an estimated 1 million species were to go extinct in short order (Sinha and Sowmiya 2019). Restoring global agriculture to a chemical-free basis would go some way in reducing the pressures on insect populations as also reducing the pesticide pollution of the rivers, seas, and oceans resulting in incalculable damage to the ecosystems of the earth. We are no longer discussing the loss of individual species (no matter how cute) here or there but of the prospect of the bottom beginning to fall out across the entire web of life globally. Our ignorance of life and life forms on earth is far more massive than our knowledge garnered in modern times, about which we seem to pride ourselves.

Objective 5

Curbing the global production of mining-based materials and energy. Apart from letting fossil fuels remain buried underground, which nature has achieved over millions of years of negentropic operation, it is also necessary to curb the production of iron and steel, cement and aluminum, both for reasons of climate change and to reduce land destruction.

Objective 6

The fastest possible deployment of renewable energy technologies, particularly solar and wind, has to be publicly supported. Other carbon-neutral energy technologies such as geo-thermal and small hydro should also be maximized, the latter because it can be linked to energy storage to stabilize intermittent sources. R & D on other promising energy technologies should also be publicly supported, though biomass-based electricity and biodiesel are not carbon-neutral. A transition toward an RE-based grid has already begun globally though it is being massively resisted by fossil fuel and nuclear industries who are responsible for delaying the grid transition which can be achieved without too much stress on consumer households.

Objective 7

Pro-nature activities to be publicly supported include a curb on the use of oceans as a dumping ground for solid, liquid, or radioactive wastes by any entity, state or private including military; activities to be aimed at revival of life in the oceans by the dredging of wastes including plastics, the revival of marine breeding grounds by appropriate treatment and protection, by efforts to protect coral reefs, by revival of dead zones adjacent to the shorelines of various countries, and by strict curbs on over-fishing; fishing communities whose livelihoods depend on the seas and oceans to be involved in this global exercise.

Objective 8

A multi-pronged strategy to reduce transportation demand globally by massive deployment of public transportation systems, many of which can be electrically driven (buses/trains/trams, etc.); by reduction of wasteful trade practices involving freight transportation particularly by air or refrigerated transportation; by making cities more bicycle and pedestrian friendly rather than automobile friendly (even if the automobiles are EVs); and R&D for hydrogen-fueled vehicles for heavy cargo transportation on land and by air, since hydrogen currently appears to be the only energy-dense fuel capable of replacing petroleum for heavy transportation.

Objective 9

"Rurbanization" of communities (that is, communities sized between urban and rural) with plenty of living and public spaces, provided with modern infrastructure and local employment requiring a minimum of daily transportation for the workforce. The concept of overcoming the "metabolic rift" between city and countryside should provide guiding principles for this reconfiguration of the population in a long-term sense. With the current level of communications technology and infrastructure, the divide between "the idiocy of rural life" and the epilepsy of post-modern urban existence can be overcome.

Objective 10

A massive increase in welfare activities and budgetary expenditures globally but especially in the developing countries to cover sectors such as food, health, education, water, energy, and physical and social infrastructure, with communities becoming more self-organized and resilient, based on non-alienated relationships and lesser hierarchical levels, permitting a greater development of human potential.

Objective 11

Much greater sharing of technology which should be treated as a common property resource in the way that the sciences are today; this requires technology to be converted to open-source knowledge rather than privatized knowledge for sustaining corporate profits.

Objective 12

Maintaining full employment along with lesser working hours per week or per year should be one of the guiding metrics in the future. Capitalist society is full of spurious material needs to cater to which there is a proliferation of spurious jobs—advertising, televangelizing, cosmetics, fake "feed-good" products, and so on—even as important needs of human communities remain unfulfilled. The list of such "socially unnecessary" products and services is very long indeed. On the other hand, society needs many more health care workers and professionals, teachers, artists and musicians, social scientists and philosophers, farmers and gardeners, animal rearers, soil specialists, and all those who will tend to the well-being of nature whether in the forests, the grasslands, the rivers, and the oceans. In effect, green socialism does not need to be tied with the present conception of alienated, joyless drudge labor performed perforce by an unhappy working class. What a person contributes through work is required by both society and nature and need not be a meaningless activity. Such a profound change in activity patterns, the very transformation of work, is inconceivable under the capitalist yoke, not even under its most advanced form of green Keynesianism.

Objective 13

Lastly, material consumption. It is possible to have highly civilized societies without excessive material consumption; yet "degrowth" is not possible within a capitalist economy because any fall in consumption leads to recession, then depression and economic crisis. It can no longer be denied that the "ecological footprint" of over-urbanized societies is responsible for both the ecological and climate crises, since rural populations even today have a much lower ecological footprint and are involved in carbon-negative or nature conserving activities. The long-term answer lies in a planned transition to a green socialist economy. Human society already has most of the technology needed for such a transition; the failure to make the transition is entirely political.

The "thought experiment" of building a green socialist economy involves both seeking out the available means as well as clarifying the end-objectives of such a transition in the near future. Most of the public discussion, particularly with respect to the climate crisis, is premised on the assumption that there is no alternative to capitalism. This has permitted capitalism to set the agenda, the framework, and boundary conditions and the time frame for dealing with the crisis. It is time to wrest the initiative away from the hands of capitalism and back into the hands of society where it rightfully belongs.

We all have the right to determine our collective future and the right to survive of future generations . . . and also of nature.

Politics is too important to be left to the politicians who promise the moon and deliver sixpence.

Chapter Seven

Summary and Conclusion

In the preceding chapters, we have tried to focus on the three real existential crises confronting humanity in the twenty-first century. We have referred to them as fast death, slow death, and certain death respectively, referring to the international arms race, the environmental crisis, and the climate crisis. A little reflection will show that the most immediately threatening crisis is that of the arms race, particularly with nuclear arms being in question. Unless these confrontational postures are firmly subdued across the world, there is no possibility of seeking global cooperation for resolving the other two crises, hence the highest priority must be accorded to its immediate and peaceful resolution. Moreover, reduction in arms expenditure globally will release ample means, budgetary and otherwise, for tackling the other two crises, which are deeply intertwined. A reduction in global arms expenditure and the chance of international peace will buy the requisite time to focus on the other two crises which will continue to haunt humanity and its future generations for centuries to come: that much is already clear from the available record.

The need to outline a model of transition toward a green socialist economy and its building blocks arises from a deeper understanding that we are already on the cusp of crisis. There is little chance of undoing what we have already done since so many of the processes unleashed are proving to be irreversible, somewhat like the Second Law of Thermodynamics which drives natural processes in a particularly direction. It is of course theoretically possible to reverse the damage resulting from the operation of the Second Law but it is far far more difficult. It is easy to burn a piece of coal or a barrel of oil, it is far more difficult to remove the CO_2 from the atmosphere and immobilize it for the requisite millennia. Similarly, it is far easier to pollute a river, far more difficult to rid the river of these pollutants or to reverse the damage to riparian life forms. From the start of the Industrial Revolution, perhaps six generations have chosen to ignore the Second Law and its ramifications in the pursuit of prosperity, power, and dominance over nature. In the process, we leave behind as a heritage for future generations a vast cleanup operation whether it be of nuclear or chemical wastes, devastated natural systems, and dismembered webs of life. In polite words, previous industrial society has followed a scorched earth policy, almost literally, with an attitude of Devil take the hindmost.

But it is today's younger generation, our own children and grandchildren, who will be inheriting this devastated planet. It is to them that this document is addressed, with a simple message from the heart: do not follow in our footsteps. Change while you can, during the brief moment of existential freedom when you step into adulthood and begin to see the world as it is.

With sorrow, we may dwell on the lines of verse written by Omar Khayyam in his Rubaiyat, nearly a thousand years ago:

> The Moving Finger Writes; and having Writ,
> Moves on; not all thy Piety nor Wit
> Shall lure it back to cancel half a Line,
> Nor all thy Tears wash out a Word of it.

References

Arendt, Hannah. (1971). *The Life of the Mind.* New York and London: Harcourt Brace Jovanovich Inc. 51–54.

Arendt, H. (1981). *The Life of the Mind.* San Diego, CA: Harcourt Brace Jovanovich.

Barlow, M., & Clarke, T. (2004). *Blue Gold.* Retrieved from https://thenewpress.com/books/blue-gold.

Beckert, S. (2014). *Empire of Cotton: A New History of Global Capitalism* (first edition). New York: Penguin.

Braudel, Fernand. (1984). *Civilization and Capitalism 15th to 18th Century—The Perspective of the World (Vol III).* English translation William Collins & Sons, London and Harper and Row, New York. 521–22.

Braudel, F. (1992). *Civilization and Capitalism, 15th–18th Century, Vol. I: The Structure of Everyday Life* (S. Reynold, Trans.). Berkeley: University of California Press.

Chomsky, N. (1970/2005). *Government in the Future.* New York: Seven Stories Press.

Commoner, B. (1976). *The Poverty of Power: Energy and the Economic Crisis* (first edition). New York: Random House.

Cook, J. (2013, July 1). Four Hiroshima Bombs Worth of Heat per Second. Retrieved from https://skepticalscience.com/4-Hiroshima-bombs-worth-of-heat-per-second.html.

Davis, P. J., & Hersh, R. (1981/1999). *The Mathematical Experience.* Boston: Houghton Mifflin.

Dhara, S. (2019, March 11). Personal communication.

Eatwell, Roger. (1996). *Fascism—A history.* London: Vintage/Random House.

Eatwell, R. (2011). *Fascism: A History.* New York: Random House.

Flannery, T. (2005/2007). *The Weather Makers: Our Changing Climate and What It Means for Life on Earth.* London: Penguin.

Foster, J. B. (2009). *The Ecological Revolution: Making Peace with the Planet.* New York: Monthly Review Press. 144, 212, 236–37, 231.

Freese, Barbara. (2003). *Coal—A human history.* New York: Perseus Publishing. Back cover, 7–8, 58–64, 105.

Freese, B. (2006). *Coal—A Human History.* London: Arrow Books.

Ghotge, S. (2018). Climate Change and Marx in the 21st Century, Part I. *Capitalism Nature Socialism, 29*(2), 29–42. https://doi.org/10.1080/10455752.2018.1467366.

Gorz, A. (2018). *Ecologica.* Retrieved from https://www.seagullbooks.org/ecologica/.

Harvey, D. (2007). *A Brief History of Neoliberalism.* Oxford: Oxford University Press.

Harvey, D. (2008). *Reading Marx's Capital with David Harvey—YouTube.* Retrieved from https://www.youtube.com/user/readingcapital.

Harvey, David. (2010). *A Companion to Marx's Capital.* London; New York: Verso.

Harvey, D. (2018). *A Companion to Marx's Capital: The Complete Edition: 1-2* (Reprint edition). London: Verso.
Hill, Christopher. (1970). *God's Englishman—Oliver Cromwell and the English Revolution.* London: Penguin Books. 243–67.
Hill, C. (2019). *God's Englishman.* New York: Penguin.
Hobsbawm, E. J. (1962). *The Age of Revolution 1789–1848.* London: Vintage Books. Preface, ix.
Hobsbawm, E. J. (1968). *Industry and Empire.* London: Pelican Books. 326, Fig 1.
Hobsbawm, E. J. (1975). *The Age of Capital 1848-1875.* London: Abacus/Little Brown. Maps 3 and 4 titled "Slavery and Serfdom in the Western World," 368–369.
Hobsbawm, E. (1988). *The Age of Capital: 1848–1875.* London: Abacus.
Hobsbawm, E. J. (1994). *The Age of Extremes—the short Twentieth Century.* London: Michael Joseph. See footnote, p. 6, Chapter 9 pp. 257–286, 43, 259; also see Chapters 1–5 for World Wars I and II, Chapter 7 for End of Empires, Chapter 8 and Chapter 9 for Cold War and Golden Years and the emergence of the military-industrial complex; 22, 31, 43, 236, 259, 261–262, 249–250.
Hobsbawm, E. (1995). *The Age of Extremes: 1914–1991.* London: Abacus.
Hobsbawm, E. (1996). *The Age of Revolution: 1789–1848* (first Vintage Books edition). New York: Vintage.
Hobsbawm, E. J. (1999). *Industry and Empire: From 1750 to the Present Day* (updated, subsequent edition; C. Wrigley, Ed.). New York: New Press.
Holmes, D. (2013). Four Hiroshima Bombs a Second: How We Imagine Climate Change. Retrieved December 27, 2019, from The Conversation website: http://theconversation.com/four-hiroshima-bombs-a-second-how-we-imagine-climate-change-16387.
IPCC. (2007a). *Climate Change 2007: Synthesis Report Summary for Policymakers.* Retrieved from IPCC website: https://www.ipcc.ch/site/assets/uploads/2018/02/ar4_syr_spm.pdf.
IPCC. (2007b). *The Physical Basis.* Retrieved from IPCC website: https://www.ipcc.ch/site/assets/uploads/2018/05/ar4_wg1_full_report-1.pdf.
Kabeer, N. (2019, March 26). India Emerged as the Third Largest Solar Market in 2018 Behind China and US. Retrieved December 28, 2019, from Mercom India website: https://mercomindia.com/india-solar-market-2018-behind-china-us/.
Landes, D. (1969). *The Unbound Prometheus.* Retrieved from https://www.abebooks.co.uk/book-search/title/the-unbound-prometheus/author/landes/.
McGuffie, K., & Henderson-Sellers, A. (2005). *A Climate Modelling Primer* (third edition). Retrieved from https://www.wiley.com/en-us/A+Climate+Modelling+Primer%2C+3rd+Edition-p-9781118687857.
Moore, Barrington. (1966). *Social Origins of Dictatorship and Democracy—Lord and Peasant in the Making of the Modern World.* London: Peregrine Books. 9, 19, 27–28, 433–42.
Moore, B. (1978). *Injustice: The Social Bases of Obedience and Revolt: The Social Bases of Obedience and Revolt.* https://doi.org/10.4324/9781315496535.
Moore, B. (1993). *Social Origins of Dictatorship and Democracy: Lord and Peasant in the Making of the Modern World.* London: Peregrine Books.
Olson, K. R., & Morton, L. W. (2019). Long-Term Fate of Agent Orange and Dioxin TCDD Contaminated Soils and Sediments in Vietnam Hotspots. *Open Journal of Soil Science, 09,* 1. https://doi.org/10.4236/ojss.2019.91001.
Patnaik, P. (2019). The Subversion of MGNREGS – IDEAs. Retrieved December 27, 2019, from http://www.networkideas.org/news-analysis/2019/02/the-subversion-of-mgnregs/.
Phillips, P. (2018). *Giants: The Global Power Elite.* New York: Seven Stories Press.
Polanyi, K. (1944/2001). *The Great Transformation: The Political and Economic Origins of Our Time* (second edition). Boston: Beacon Press.
Prud'homme, A. (2011). *The Ripple Effect: The Fate of Freshwater in the Twenty-First Century* (reprint edition). New York: Scribner.
Raju, S., & Ramana, M. V. Cost of Electricity from the Jaitapur Nuclear Power Plant; *Economic and Political Weekly, 48*(26 and 27, 29), 58.
Raju, S., & Ramana, M. V. (2015). Cost of Electricity from the Jaitapur Nuclear Power Plant. *Economic and Political Weekly, 48*(26–27), 7–8.

Ricke, K. L., & Caldeira, K. (2014). Maximum Warming Occurs about One Decade after a Carbon Dioxide Emission. *ERL*, *9*(12), 124002. https://doi.org/10.1088/1748-9326/9/12/124002.

Shirer, William L. (1950). *The Rise and Fall of the Third Reich*. New York: Fawcett Crest. 30th Anniversary Edition, 569–77, 638–46, 655–64, 668–77, 700–5, 1062.

Shirer, W. L. (1960). *The Rise and Fall of the Third Reich* (first edition). Fawcett Crest.

Sinha, A., & Ashok, S. (2019, May 8). One Million Species Face Extinction: Why Biodiversity Report Matters. *The Indian Express*. Retrieved from https://indianexpress.com/article/explained/simply-put-why-biodiversity-report-matters-5715922/.

Streeck, W. (2014). *This Time Is Serious: Capitalism on the Brink*. Retrieved from https://mediacentral.princeton.edu/media/This+Time+is+SeriousA++Captialism+on+the+Brink/1_hjtxu54u.

Thomson, D. (1974). *Europe Since Napoleon*. Jaipur, India: Jain Pustak Mandir.

Thurow, L. C. (1980). *Zero-sum Society*. New York: Basic Books.

Watts, J. S. (2010). *When A Billion Chinese Jump: How China Will Save Mankind—Or Destroy It*. New York: Simon and Schuster.

Weizsacker, E. von, Lovins, A. B., & Lovins, L. H. (1998). *Factor Four: Doubling Wealth, Halving Resource Use—The New Report to the Club of Rome* (first edition). London: Routledge.

Weizsacker, E. U. von, Hargroves, C., Smith, M. H., Desha, C., & Stasinopoulos, P. (2009). *Factor Five: Transforming the Global Economy through 80% Improvements in Resource Productivity* (first edition). London; Sterling, VA: Routledge.

Index

About the Author

Sanjeev Ghotge was previously a professor at the Centre for Applied Systems Analysis in Development at Pune. Later he was a senior fellow and head of the Centre for Climate and Sustainability Policy at the World Institute of Sustainable Energy, Pune. He is now retired and lives in Pune, India.

www.ingramcontent.com/pod-product-compliance
Lightning Source LLC
LaVergne TN
LVHW091533200926
848483LV00002B/390
* 9 7 8 1 4 9 8 5 9 5 7 3 5 *